T0320527

Introduction to Topological Quantum Computation

Combining physics, mathematics and computer science, topological quantum computation is a rapidly expanding research area focused on the exploration of quantum evolutions that are immune to errors. In this book, the author presents a variety of different topics developed together for the first time, forming an excellent introduction to topological quantum computation.

The makings of topological systems, their properties and their computational power are presented in a pedagogical way. Relevant calculations are fully explained, and numerous worked examples and exercises support and aid understanding. Special emphasis is given to the motivation and physical intuition behind every mathematical concept.

Demystifying difficult topics by using accessible language, this book has broad appeal and is ideal for graduate students and researchers from various disciplines who want to get into this new and exciting research field.

Jiannis K. Pachos is a Reader in the School of Physics and Astronomy at the University of Leeds, UK. He works on a variety of research topics, ranging from quantum field theory to quantum optics. Dr Pachos is a University Research Fellow of the Royal Society.

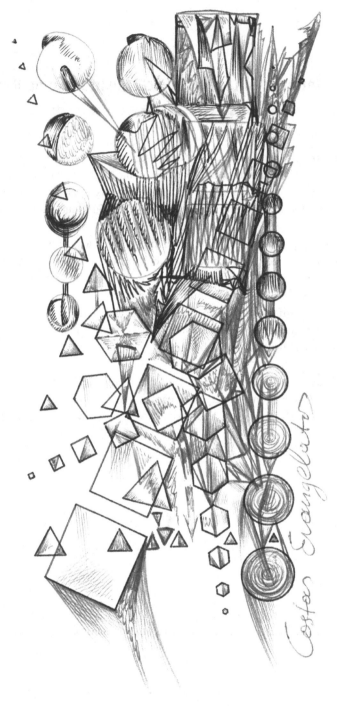

Introduction to Topological
Quantum Computation

JIANNIS K. PACHOS

University of Leeds, UK

CAMBRIDGE
UNIVERSITY PRESS

CAMBRIDGE
UNIVERSITY PRESS

University Printing House, Cambridge CB2 8BS, United Kingdom

Cambridge University Press is part of the University of Cambridge.

It furthers the University's mission by disseminating knowledge in the pursuit of education, learning and research at the highest international levels of excellence.

www.cambridge.org
Information on this title: www.cambridge.org/9781107005044

© J. K. Pachos 2012

First published 2012

A catalogue record for this publication is available from the British Library

ISBN 978-1-107-00504-4 Hardback

To Almut and Sevi

Contents

Acknowledgements

I am grateful to Almut Beige and Ville Lahtinen for their continuous support and guidance during the writing of this book. I would also like to thank several people for their direct or indirect support, such as Miguel Aguado, Abbas Al-Shimary, Gavin Brennen, Michael Freedman, Silvano Garnerone, Sofyan Iblisdir, Roman Jackiw, Petr Jizba, Louis Kauffman, Nikolai Kiesel, Alexei Kitaev, Lauri Lehman, Samuel Lomonaco, Mark Mitchison, David Perez-Garcia, So-Young Pi, John Preskill, Nicholas Read, Renato Renner, Emily Riley, Christian Schmid, Ady Stern, David Tong, Zhenghan Wang, Harald Weinfurter, Witlef Wieczorek, James Wootton, Paolo Zanardi and Vaclav Zatloukal.

PART I

PRELIMINARIES

1   Introduction

Symmetries play a central role in physics. They dictate what one can change in a physical system without affecting any of its properties. You might have encountered symmetries like translational symmetry, where a system remains unchanged if it is spatially translated by an arbitrary distance. A system with rotational symmetry, however, is invariant under rotations. Some symmetries, like the ones mentioned above, give information about the structure of the system. Others have to do with the more fundamental physical framework that we adopt. An example for this is the invariance under Lorentz transformations in relativistic physics.

Other types of symmetries can be even more subtle. For example, it is rather self-evident that physics should remain unchanged if we exchange two identical point-like particles. Nevertheless, this fundamental property that we call statistical symmetry gives rise to rich and beautiful physics. In three spatial dimensions it dictates the existence of bosons and fermions. These are particles with very different quantum mechanical properties. Their wave function acquires a $+1$ or a -1 phase, respectively, whenever two particles are exchanged. A direct consequence of this is that bosons can actually occupy the same state. In contrast, fermions can only be stacked together with each particle occupying a different state.

When one considers two spatial dimensions, a wide variety of statistical behaviours is possible. Apart from bosonic and fermionic behaviours, arbitrary phase factors, or even non-trivial unitary evolutions, can be obtained when two particles are exchanged (Leinaas and Myrheim, 1977). Particles with such exotic statistics have been named anyons by Frank Wilczek (1982). The transformation of the anyonic wave function is consistent with the exchange symmetry. Indeed, similarly to the fermionic case, the anyonic exchange transformations are not detectable by local measurements on the particles. This 'indirect' nature of the statistical transformations of anyons is at the core of their intellectual appeal. It also provides the technological advantage of anyonic systems in performing quantum computation that is protected from a malicious environment.

1.1 Particle exchange and quantum physics

Statistics, as arising from indistinguishability of particles, is a quantum mechanical property. Classical particles are always distinguishable as we can keep track of their position at all times. Quantum mechanically, the position of a particle is determined via a spatially extended wave function. The wave functions of two particles might overlap even if they are not peaked at exactly the same position. Hence the position is, in general, not a good property for identifying particles, thereby making it impossible to define distinguishability in a fundamental way. This suggests adopting a common wave function to describe the system of the two particles.

Indistinguishable particles in quantum mechanics should have all their intrinsic properties, such as mass, charge, spin and any other quantum number, exactly the same. This seemingly innocent property has far-reaching consequences. It allows us to construct universal theories to describe elementary particles based on simple statistical rules. More dramatically, it forces us to adopt the new framework of statistical physics that abandons the distinguishability of particles.

Exchange statistics describe the change in the wave function of two identical particles, when they are exchanged. Its properties need to be compatible with the symmetry imposed by indistinguishability. As an important consequence, these changes are independent of many details of the system. Consider, for example, the case where the exchange is not a mathematical procedure, but a physical process of moving two particles along an exchange path. The effect of this transport on the wave function should not depend on the particular shape of the path taken by the particles when they are exchanged or the speed the path is traversed. Nevertheless, the evolution might still depend on some global, topological characteristics of the path, such as the number of times the particles are exchanged. Statistical evolutions are hence topological in their nature.

In three spatial dimensions the indistinguishability of particles allows for the possibility of having bosons and fermions. Bosons satisfy the Bose–Einstein distribution (Bose, 1924; Einstein, 1924) and fermions the Fermi–Dirac distribution (Fermi, 1926; Dirac, 1926). These distributions emerge from the general requirement that an ensemble of indistinguishable particles is described either by completely symmetric or completely antisymmetric wave functions with respect to particle exchanges. The first case corresponds to bosons and the second to fermions. In particular, when two fermions are positioned on top of each other their state should be simultaneously symmetric and antisymmetric, giving zero as the only possible solution. This gives rise to the Pauli exclusion principle that assigns zero probability to such configurations. However, there is no such restriction for the case of bosons which can freely occupy the same position.

Another surprising consequence of indistinguishability is the relation between spin and statistics. Pauli (1940) proved that bosons have integer spin and fermions half-integer spin. This is a rather surprising relation as spin is an intrinsic property that can be determined by considering an isolated particle. Contrarily, to determine the statistics we need to consider an ensemble of at least two particles. We shall visit this relation again later on and we shall generalise it to the case of anyons where exotic statistical behaviours give rise to equally exotic values of spins.

1.2 Anyons and topological systems

Statistics is spectacularly manifested in two-dimensional systems. There, exotic wave functions of particles can be realised that give rise to anyons. The study of anyons started as a theoretical curiosity in two-dimensional models (Wilczek, 1982). However, it was soon realised that they can be encountered in physical systems with effective two-dimensional behaviour. For example, gases of electrons confined on thin films in the presence of sufficiently strong magnetic field and at a sufficiently low temperature give rise to the fractional quantum Hall effect (Camino *et al.*, 2005; Laughlin, 1983; Tsui *et al.*, 1982). The low-energy excitations of these systems are localised quasiparticle excitations that exhibit anyonic statistics. Beyond the fractional quantum Hall effect, other two-dimensional systems have emerged which theoretically support anyons (Volovik, 2003). These range from superconductors (Chamon *et al.*, 2001) and topological insulators (Hasan and Kane, 2010) to spin lattice models.

Systems that support anyons are called topological as they inherit the topological properties of the anyonic statistical evolutions. Topological systems are usually many-particle systems that support localised excitations, so-called quasiparticles, that can exhibit anyonic behaviour. In general, they have highly entangled degenerate ground states. As a consequence local order parameters, such as the magnetisation, are not able to describe topological phases. So we need to employ non-local order parameters. Various characteristics exist that identify topological order, such as ground state degeneracy, edge states in the presence of a gapped bulk, topological entanglement entropy or the explicit detection of anyons. As topological order comes in various forms, the study and characterisation of topological systems in their generality is complex and still an open problem. Over the last years the richness in the behaviour of two-dimensional topological systems has inspired many scientists. One of the most thought-provoking ideas is to use topological systems for quantum computation.

1.3 Quantum computation with anyons

In the last decades progress in physics and the understanding of nature has advanced the way we perceive information. Quantum physics has opened the possibility of yet another way of storing, manipulating and transmitting information. Importantly, quantum computers have been proposed with the ability to outperform their classical counterparts, thereby promising far-reaching consequences. Quantum computation requires the encoding of quantum information and its efficient manipulation with quantum gates (Nielsen and Chuang, 2000). Qubits, the quantum version of classical bits, provide an elementary encoding space. Quantum gates manipulate the qubits to eventually perform a computation. A universal quantum computer employs a sufficiently large set of gates in order to perform arbitrary quantum algorithms. In recent years, there have been two main quests for quantum computation. First, to find new algorithms, that go beyond the already discovered algorithms of searching (Grover, 1996) and factorising (Shor, 1997). Second, to perform quantum computation that is resilient to errors.

In the 1990s a surprising connection was made. It was argued by Castagnoli and Rasetti (1993) that anyons could be employed to perform quantum computation. Kitaev (2003) demonstrated that anyons could actually be used to perform fault-tolerant quantum computation. This was a very welcome advance as errors infest any physical realisation of quantum computation, coming from the environment or from control imperfections. Shor (1995) and Steane (1996) independently demonstrated that for sufficiently isolated quantum systems and for sufficiently precise quantum gates, quantum error correction can allow fault-tolerant computation. However, the required thresholds are too stringent and demand a large overhead in qubits and quantum gates for error correction to be realised. In contrast to this, anyonic quantum computation promises to resolve the problem of errors from the hardware level.

Topological systems can serve as quantum memories or as quantum computers. They can encode information in a way that is protected from environmental perturbations. In fact, topological systems have already proven to be a serious candidate for constructing fault-tolerant quantum hard disks. The intertwining of anyons and quantum information in topological systems is performed in an unusual way. Information is encoded in the possible outcomes when bringing two anyons together. This information is not accessible when the anyons are kept apart, and hence it is protected. The exchange of anyons gives rise to statistical logical gates. In this way anyons can manipulate information with very accurate quantum gates, while keeping the information hidden at all times. If the statistical evolutions are complex enough then they can realise arbitrary quantum algorithms. Fundamental properties of anyonic quasiparticles can thus become the means to perform quantum computation. Fault-tolerance simply stems from the ability to keep these quasiparticles intact. The result is a surprisingly effective and aesthetically appealing method for performing fault-tolerant quantum computation.

1.4   Abelian and non-Abelian anyonic statistics

It is commonly accepted that in three spatial dimensions indistinguishable particles, elementary or not, come in two species: bosons or fermions. The possibility for these statistical behaviours can be obtained from a simple thought experiment. Consider two identical particles in three dimensions, where one of them circulates the other via the path C_1, as shown in Figure 1.1(a). As we are only interested in the statistical behaviour of these particles, we focus on the topological characteristics of this process. These characteristics should be independent of details such as the particular geometry of the path or direct interactions between the particles. Hence, we can continuously deform the path C_1 to the path C_2. This involves only local deformations of the evolution without cutting or otherwise drastically changing the nature of the path. In its turn, path C_2 can be continuously deformed to a trivial path, C_0, that keeps the particle at its initial position at all times. As a consequence, the wave function, $\Psi(C_1)$, of the system after the circulation has to be exactly the same as the original one $\Psi(C_0)$, i.e.,

$$\Psi(C_1) = \Psi(C_2) = \Psi(C_0). \tag{1.1}$$

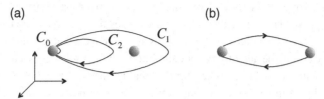

(a) A particle spans a loop around another one. In three dimensions it is always possible to continuously deform the path C_1 to the path C_2, which is equivalent to a trivial path, C_0. (b) Two successive exchanges between two particles are equivalent to a circulation of one particle around the other and a translation.

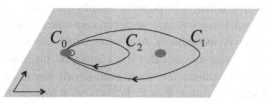

In two dimensions the two paths C_1 and C_2 are topologically distinct. This gives the possibility of having non-trivial phase factors appearing when one particle circulates around the other. In other words, one can assign a non-trivial unitary to the evolution corresponding to path C_1.

Figure 1.1(b) depicts a single exchange of two particles. If we perform two of these exchanges in succession then we obtain a full circulation of one particle around the other accompanied by an irrelevant spatial translation. Thus, a single exchange can result in a phase factor $e^{i\varphi}$ that has to square to unity in order to be consistent with (1.1). This has two solutions, $\varphi = 0$ and $\varphi = \pi$, corresponding to the bosonic and fermionic statistics, respectively. These are the only statistical behaviours that can exist in three spatial dimensions.

When we restrict ourselves to two spatial dimensions, then we are faced with a wealth of possible statistical evolutions. If the particle circulation C_1 is performed on a plane, as shown in Figure 1.2, then it is not possible to continuously deform it to the path C_2, as we do not have access to an extra dimension to lift the loop and undo the linking. To do that would necessitate cutting the path, passing it over the circulated particle and glueing it again, thus changing in-between its topological characteristics. Still, the evolution that corresponds to C_2 is equivalent to the trivial evolution. As we are not able to deform the evolution of path C_1 to the trivial one, the argument we employed in the three-dimensional case does not apply any more. Actually, now, it is possible to assign an arbitrary phase factor, or even a whole unitary matrix, to the evolution corresponding to C_1 that is equivalent to two successive exchanges. Thus, particles in two dimensions can have rich statistical behaviours.

We would like now to analyse the difference between phase factors and unitary matrices as statistical evolutions. In the former case the anyons are known as Abelian, and the statistical phase factor can take any value between the bosonic case of $e^{i\varphi_b} = 1$ and the fermionic case of $e^{i\varphi_f} = -1$. In fact, it is the possibility of these particles having any statistics that led to the name 'anyon' (Wilczek, 1982). Nevertheless, the statistics of a specific anyon type is always well defined and a given pair will always yield the same statistical phase. This phase therefore characterises the species of the exchanged anyons.

Beyond a phase factor it is possible to have a statistical evolution that is more complex. Certain species of anyons called non-Abelian give rise to an exchange evolution that can actually lead to a higher-dimensional unitary matrix. In contrast to phase factors, matrices do not in general commute, which motivates the name 'non-Abelian'. For a matrix statistical evolution to emerge, the wave function that describes the particles needs to be part of a degenerate subspace of states. The particle exchange then transforms between states in this subspace without changing the energy of the system. Nevertheless, there is an important constraint we need to impose on the statistical evolution in order to be in agreement with the exchange symmetry. To preserve the physics when two identical non-Abelian anyons are exchanged, we require that the degenerate states should be non-distinguishable if one looks at each anyon individually. As a result, interchanging these anyons causes a transformation within this state subspace that is not detectable by local measurements, giving a valid statistical transformation. One would need to perform non-local operations, like bringing these anyons close together, in order to distinguish between these states and observe the effect of statistics. It is rather surprising that consistent particle theories exist that have such exotic behaviours as the non-Abelian statistics. Before characterising these theories we shall first investigate the physical principles that allow this behaviour to emerge.

1.5   What are anyonic systems?

The study of anyons becomes even more exciting when the possibility arises to realise them in the laboratory. To date it is believed that Abelian anyons have already been detected in the laboratory (Camino *et al.*, 2005), and there is strong evidence for the existence of non-Abelian anyons (Willett *et al.*, 2009). But how is it possible to construct a purely two-dimensional world, where the exotic properties of anyons can emerge? In order to determine how plausible this is, we need to identify the main characteristics of anyons. Only then can we decide whether we can physically realise topological systems that can support anyons.

1.5.1 Two-dimensional wave functions and quasiparticles

Admittedly, our physical world appears to be three- and not two-dimensional. This is also well manifested in the statistical properties of the elementary particles accounted for in nature, bosons and fermions. The natural question then arises: how is it possible to obtain a two-dimensional world where anyons can emerge? Even if we make a system arbitrarily thin, it is impossible to trick nature into believing that it is actually reduced to two dimensions. To the rescue comes quantum mechanics. It is possible to construct a quantum system with a wave function that splits, via the separation of variables method, to a purely two-dimensional and a one-dimensional part. Let us analyse this in more detail.

To determine the behaviour of a particle in three spatial dimensions with position $\mathbf{r} = (x, y, z)$, subject to a potential of the form

$$V(\mathbf{r}) = V_{xy}(x, y) + V_z(z), \qquad (1.2)$$

we can employ the separation of variables method. In this case the wave function can be written as

$$\Psi(\mathbf{r}) = \Psi_{xy}(x, y)\Psi_z(z), \qquad (1.3)$$

where $\Psi_{xy}(x, y)$ satisfies the two-dimensional Schrödinger equation subject to the potential $V_{xy}(x, y)$ and $\Psi_z(z)$ satisfies a one-dimensional Schrödinger equation subject to $V_z(z)$. Hence, the wave function $\Psi_{xy}(x, y)$ is purely two-dimensional with its dynamics decoupled from the third direction z.

Consider now the system being homogeneously confined along the z direction. The low energy levels corresponding to this trapping are discrete. For strongly confining potentials the typical energy splitting, ΔE, between these levels is large. Let us take the particle to be initially prepared in the ground state. If the particle is subject to additional dynamics like a perturbation, beyond the trapping potential, with a scale much smaller than ΔE, then the particle will remain in the same energy level. This is an important mechanism for reducing the dimensionality of the system from three to two by suppressing the motion in the third direction. We also demand the presence of an energy gap that separates $\Psi_{xy}(x, y)$ from the two-dimensional excited states. This gap protects the characteristics of $\Psi_{xy}(x, y)$ against external perturbations. Under these conditions the behaviour of the system is essentially given by the two-dimensional wave function $\Psi_{xy}(x, y)$.

It is important to notice that the finite energy scales that either isolate the anyonic behaviour of the reduced state $\Psi_{xy}(x, y)$ from spurious excitations or suppress the motion in the third direction are the Achilles' heel that makes anyonic systems fragile. Indeed, when perturbations or temperature are strong enough compared to these energy scales then either the anyonic characteristics are washed out or the state of the system stops being two-dimensional. Hence, we need to keep track of such spurious effects in order to ensure reliable anyonic behaviour. Needless to say, if we had a truly two-dimensional system then anyons would be fundamental particles and they would be robust even at much higher energies. This sensitivity of effective anyonic models is a main challenge for topological quantum computation.

The particles that are subject to the above conditions do not actually see only two dimensions, but their wave function becomes effectively two-dimensional. Hence, we cannot expect the constituent particles to automatically acquire anyonic properties. Nevertheless, we could expect that effective particles, so-called quasiparticles, emerge from the properties of many-particle wave functions that are truly two-dimensional. In Figure 1.3 a many-particle system is shown and the possible emergence of quasiparticles is described.

Quasiparticles are entities defined through the wave function of a many-particle system. They behave like particles, i.e., they have local properties and they respond to their local

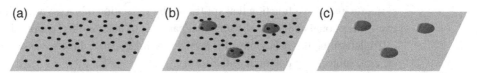

Fig. 1.3 (a) A system with constituent particles confined on a plane that give rise to a two-dimensional wave function. (b) Quasiparticles are identified as localised properties of the two-dimensional wave function of the constituent particles. (c) Often we forget the constituent particles and we treat the quasiparticles as elementary ones living on the two-dimensional space.

environment. Such a behaviour emerges, for example, when the constituent particles of the system interact in such a way that they give rise to exotic, highly correlated wave functions. Importantly, quasiparticles can have properties that are completely different from the properties of the constituent particles. One could expect that anyonic properties could emerge in this way. This is indeed the case for all known examples of topological systems, ranging from the fractional quantum Hall effect to spin lattice models that exhibit topological behaviour. Hence, the search for anyons becomes intrinsically related to the study of strongly correlated quantum mechanical systems.

Another aspect of the quasiparticle nature of anyons is that all anyons emerge from the same wave function of the whole system. They are aware of each other's position, which makes it possible to exhibit the desired exchange statistics. More concretely, the exchange statistics emerges as an evolution of this wave function that depends on the history of the constituent particles. Indeed, we shall see in the next chapter how the statistical evolutions of anyons can manifest themselves as geometric phases of the global wave function of the system.

1.5.2 Symmetry, degeneracy and quantum correlations

From the previous subsection it becomes apparent that anyons, emerging as quasiparticle states of a many-particle system, are purely quantum mechanical objects. Now we would like to discuss how strongly correlated these topological systems need to be in order to give rise to anyonic properties. To be concrete we consider two anyonic properties which are tightly connected to quantum correlations of the constituent particles. First, we intuitively approach the invariance of statistical evolutions in terms of continuous deformations of the paths used to exchange the quasiparticles. Second, we analyse the non-Abelian character which is manifested as an evolution acting on degenerate ground states.

The statistical transformation occurring under exchange of quasiparticles should be the same for arbitrary shapes of the path chosen for the exchange, as long as they can be continuously deformed into each other. This is an important property that gives rise to the topological character of statistical evolutions. It is equivalent to requiring invariance of the evolution when the coordinates of the system are continuously deformed, or in other words when the spanned paths are continuously deformed. The transport of quasiparticles can be described by products of local operators that act on the states of the constituent particles of the system. In order to have evolutions that are invariant under continuous deformations of the paths, the states of the system need to be invariant under the action of particular local configurations of such operators. The set of all possible path deformations is large, giving rise to an equivalently large set of operators that leave invariant the states of the system. Apart from the case where these operators are trivial, such highly symmetric states are highly correlated. We shall meet such examples in the following chapters.

Consider now the case where the system exhibits non-Abelian statistics. In this case the statistical evolution is a unitary matrix that acts on a state space, whose states should all have the same energy. Otherwise, transforming between them along an exchange would not correspond to a statistical symmetry. Hence, a degenerate subspace is required to encode such statistical evolutions. In order for the corresponding evolution to be statistical, these states should not be distinguishable in any local way, i.e., no obvious local characteristic should exist that witnesses the statistical evolution. In other words, statistical symmetry imposes that local observables should remain invariant under statistical evolutions.

How can a degeneracy between locally indistinguishable states give information about correlations of the system? It is understood that a local symmetry creates a degeneracy in the system. The distinction between such degenerate states can be observed with a local operator. Therefore, the local symmetry mechanism cannot be responsible for creating the degeneracy of non-Abelian anyonic systems. It is known that strongly correlated systems often exhibit degeneracy in the ground state that does not correspond to local symmetries. Such strongly correlated systems are prime candidates for creating the degeneracy required by a topological system that supports non-Abelian statistics.

Summary

In this chapter we introduced the concept of particle statistics. Very simple principles restrict the statistical behaviour of particles in three spatial dimensions to be only bosonic or fermionic. In two dimensions particles are allowed to exhibit exotic statistics. These can be described by phase factors or whole unitary matrices instead of a plus or a minus sign that corresponds to bosons or fermions, respectively. These exotic particles are called anyons.

We want to employ the statistical evolutions of anyons as a novel way to perform quantum computation. This promises to efficiently overcome the problem of errors that prohibit the reliable storing and manipulation of quantum information. Employing anyons for such technological tasks requires a good understanding of their properties. We also need to investigate in detail the properties of the topological systems that support anyons.

Anyons are expected to emerge as localised properties, so-called quasiparticles, in the wave function of many-particle systems. We argued that these wave functions need to be highly correlated if they are to support anyonic statistics. The study of highly correlated quantum systems that can support anyons is the topic of subsequent chapters in this book.

Exercises

1.1 By considering the wave function $\Psi(\mathbf{r}_1, \mathbf{r}_2)$ of two fermions at positions $\mathbf{r}_1$ and $\mathbf{r}_2$, show explicitly that the Fermi statistics and the Pauli exclusion principle are compatible.

1.2 In two spatial dimensions it is possible to construct non-trivial topological configurations from a point and a looping string. For example, the configuration of a string enclosing the point is topologically non-equivalent to a loop that does not enclose it, as shown in Figure 1.2. What type of geometrical objects do we need in order to construct topological configurations in one, two, three and four dimensions?

1.3 Consider a potential of the form

$$V(\mathbf{r}) = V_{xy}(x, y) + V_z(z) + g(y, z). \tag{1.4}$$

Treating $g(y, z)$ as a small or a large perturbation compared to both $V_{xy}(xy)$ and $V_z(z)$, show when the separation of variables method breaks down, precluding the dimensional reduction of the corresponding wave function.

2

Geometric and topological phases

Anyonic statistics is manifested by phase factors resulting from moving two anyons around each other. This physical process closely resembles the Aharonov–Bohm effect (Aharonov and Bohm, 1959). There, the wave function of a charged particle acquires a phase factor when it circulates a magnetic flux confined in an infinite solenoid. This phase does not depend on the details of the traversed path, but only on the number of times the particle circulates the solenoid. A similar topological effect is also present in the statistical evolutions of particles. We shall see below that the analogy between anyons and the Aharonov–Bohm effect can be made rigorous. Still, realising anyons with actual magnetic fluxes and electric charges is not very appealing as it requires building complex mechanical structures. It is intriguing that effective fluxes and charges can arise in highly correlated systems. The best way to describe the interaction between these effective fluxes and charges is through geometric phases, also known as Berry phases (Berry, 1984). These phases provide the natural mechanism that gives rise to the anyonic statistics in many-body quantum systems.

Classical mechanics, including electrodynamics, can be cast purely in terms of real numbers. Quantum mechanics intrinsically incorporates complex numbers. The Schrödinger equation

$$i\hbar\frac{\partial\Psi}{\partial t} = H\Psi \tag{2.1}$$

has an imaginary number in front of the time derivative. Hence, its solutions Ψ are in general complex (Hardy, 2001). However, determining physical quantities concerned with the full system requires only absolute values. In this case the knowledge of any phase factors of Ψ is obsolete. Complex phases become important when considering the evolution of parts of the system. This is the case in interference experiments between different parts, which can determine their relative phases.

Typical examples where complex quantum phases appear are the Aharonov–Bohm effect (Aharonov and Bohm, 1959) and Berry phases (Berry, 1984). Moreover, quantum phases are at the heart of some of the most surprising effects of quantum physics, like the double-slit experiment (Feynman, 1965), the adiabatic approximation (Born and Fock, 1928) and Anderson localisation (Anderson, 1958).

In this chapter we investigate the emergence of quantum phases within the context of gauge fields and of geometric evolutions. This will help us to approach the physics of anyons intuitively. We shall study the geometric phases in some detail, as many of their properties carry forward to the case of braiding evolutions of anyons. Finally, we shall introduce the main mechanism behind the integer quantum Hall effect which relates to the physics of both the gauge fields and of the geometric phases.

2.1 Quantum phases from gauge fields

When a charged particle is moving in the presence of a gauge field its wave function can acquire a quantum phase. The Aharonov–Bohm effect (Aharonov and Bohm, 1959) describes such a phase shift when the particle circulates a magnetic flux tube inaccessible by the particle.

2.1.1 Charged particle in a magnetic field

A magnetic field can be described by a vector potential $\mathbf{A} = (A_x, A_y, A_z)$ via

$$\mathbf{B} = \nabla \times \mathbf{A}. \tag{2.2}$$

Any gradient of a scalar function, ω, can be added to $\mathbf{A}$ without changing the value of the magnetic field,

$$\mathbf{B} = \nabla \times (\mathbf{A} + \nabla\omega) = \nabla \times \mathbf{A}, \tag{2.3}$$

as $\nabla \times \nabla\omega = 0$ identically. The invariance of the magnetic field under different choices of ω is called gauge invariance and $\mathbf{A}$ is also known as a gauge field. Consider now a particle of charge q at position $\mathbf{r} = (x, y, z)$ moving along a looping trajectory in the presence of a magnetic field, as shown in Figure 2.1(a). The non-relativistic Hamiltonian of this system is given by the minimal coupling prescription

$$H^A = -\frac{\hbar^2}{2m}\left(\nabla - i\frac{q}{c\hbar}\mathbf{A}\right)^2. \tag{2.4}$$

It is possible to check that if $\Psi(\mathbf{r})$ is an eigenstate of this Hamiltonian with $\mathbf{A} = 0$ then the eigenstate with the same energy for a general vector potential $\mathbf{A} \neq 0$ is given by

$$\Psi^A(\mathbf{r}) = \exp\left(i\frac{q}{c\hbar}\int_{\mathbf{r}_0}^{\mathbf{r}} \mathbf{A}(\mathbf{r}') \cdot d\mathbf{r}'\right)\Psi(\mathbf{r}), \tag{2.5}$$

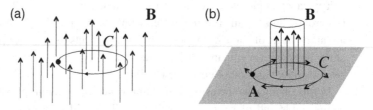

Fig. 2.1 (a) A charged particle traverses a loop C in the presence of a magnetic field **B**. The wave function of the particle acquires a phase factor that is proportional to the flux passing through a surface enclosed by the loop. (b) When the magnetic field **B** is confined inside an impenetrable solenoid, having zero value outside it, the phase factor depends only on the number of times the particle circulates the solenoid. The vector potential **A** along the path is also depicted.

where $\mathbf{r}_0$ is an arbitrary reference point and the integral is along a path connecting $\mathbf{r}_0$ and $\mathbf{r}$. The eigenstate $\Psi^A(\mathbf{r})$ shows explicitly the relation between phase factors and gauge fields. Assume that the particle is adiabatically moved (Messiah, 1962) in a looping trajectory C and we ignore any other effect apart from the interaction between the charge, q, of the particle and the magnetic field, $\mathbf{B}$. Then the wave function at the end of the cyclic evolution acquires the phase

$$\varphi = \frac{q}{c\hbar} \oint_C \mathbf{A} \cdot d\mathbf{r}. \tag{2.6}$$

By employing Stokes' theorem this phase can be written as

$$\varphi = \frac{q}{c\hbar} \iint_{S(C)} \nabla \times \mathbf{A} \cdot d\mathbf{s} = \frac{q}{c\hbar} \iint_{S(C)} \mathbf{B} \cdot d\mathbf{s} = \frac{q}{c\hbar} \Phi. \tag{2.7}$$

Here $d\mathbf{r}$ is an elementary segment of the loop C, $S(C)$ is a surface enclosed by C and $d\mathbf{s}$ is a surface element of $S(C)$, while Φ is the flux of the magnetic field that goes through $S(C)$. This phase is gauge-invariant, i.e., it does not depend on the choice of $\mathbf{A}$ provided that it gives the same magnetic field $\mathbf{B}$. Moreover, the wave function does not change if we add a unit flux $\Phi_0 = hc/q$ to the system that gives $\varphi = 2\pi$. It is apparent that the phase φ acquired by the wave function due to the looping trajectory depends exclusively on the geometry of the loop C. Nevertheless, it is invariant under deformations of C that keep the corresponding flux Φ fixed.

2.1.2 The Aharonov–Bohm effect

To describe the Aharonov–Bohm effect, let us consider the setting where magnetic flux is confined in an infinite impenetrable tube. This flux can be produced, e.g., from a series of magnetic dipoles aligned along the solenoid. We take a charged particle to move on a plane perpendicular to the tube, as shown in Figure 2.1(b). As we shall see below the charged particle can acquire a phase factor, even if it is moving in an area free from any electromagnetic field.

In particular, we take a magnetic field that is confined inside an infinitesimally thin solenoid with finite flux Φ going through it. If we position the solenoid at the origin of the (x, y, z) coordinates, the corresponding vector potential is given by

$$\mathbf{A}(\mathbf{r}) = \left(-\frac{y\Phi}{2\pi r^2}, \frac{x\Phi}{2\pi r^2}, 0 \right), \tag{2.8}$$

where $r = |\mathbf{r}|$. This corresponds to the magnetic field

$$\mathbf{B}(\mathbf{r}) = \nabla \times \mathbf{A}(\mathbf{r}) = \hat{\mathbf{z}}\Phi\delta(r). \tag{2.9}$$

Hence the magnetic field is zero at $\mathbf{r} \neq 0$, i.e., outside the solenoid, while it gives rise to a non-zero flux Φ due to its singular behaviour at $\mathbf{r} = 0$. The vector potential (2.8) is parallel to the plane with a non-trivial circulation $\oint_C \mathbf{A} \cdot d\mathbf{r}$ along any closed path C that goes around the solenoid, as shown in Figure 2.1(b). As equation (2.5) still applies, we can expect to acquire a non-trivial phase factor when the charge particle circulates the solenoid. Indeed, the phase factor is given by (2.7), where now Φ is the flux confined in the solenoid. Importantly, this is a topological effect as the phase is completely independent of the detailed shape of the path. It is only proportional to the number of times the particle circulates the solenoid. Although the particle does not interact directly with the magnetic field, its wave function responds to the presence of a non-zero vector potential. The latter mediates the information of the magnetic flux over arbitrarily long distances, thus giving rise to a non-trivial phase factor. This unexpected shift of the wave function, which became known as the Aharonov–Bohm effect, has been detected experimentally (Peshkin and Tonomura, 1989).

2.1.3 Anyons and Aharonov–Bohm effect

By drawing the analogy to the Aharonov–Bohm effect, we can obtain a somewhat mechanical picture of anyonic behaviour. Consider two composite particles equipped with a charge, q, and a magnetic dipole that gives rise to the flux, Φ, as shown in Figure 2.2(a). For simplicity we set, in the following, $c = 1$ and $\hbar = 1$. Circulating anyon 1 around anyon 2, the charge of 1 goes around the flux of 2, thereby giving rise to the Aharonov–Bohm effect with phase factor $e^{iq\Phi}$. Similarly, the magnetic dipole that gives the flux of 1 goes around the charge of 2. It equivalently gives the same phase factor $e^{iq\Phi}$. Hence, the total contribution to the wave function of the two particles is $e^{2iq\Phi}$. This phase does not depend on the details of the path, such as its shape or the rate it is spanned, as long as the adiabaticity condition is satisfied. It depends only on the number of times one particle circulates around the other. So it is topological in nature and it can faithfully describe the mutual statistics of the particles. The statistical angle of these anyonic particles, which corresponds to the phase shift of their wave function when they are exchanged, is thus given by

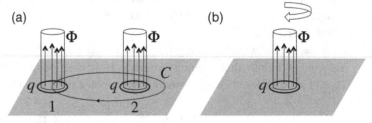

Fig. 2.2   (a) Anyons can be described effectively as composite particles with an attached magnetic flux, Φ, and a ring with electric charge, q. When anyon 1 moves around anyon 2 along loop C its charge circulates the flux of the other anyon and the Aharonov–Bohm effect gives rise to a non-trivial statistical phase. (b) When the composite particle rotates around itself by 2π it again acquires a phase factor as its charge circulates its flux, which can be attributed to a spin.

$$\varphi = q\Phi. \tag{2.10}$$

This description also encodes the spin characteristic of anyons. A 2π rotation of an anyon around itself, as shown in Figure 2.2(b), gives rise to the phase $e^{iq\Phi}$ due to the charged ring circulating the confined flux. In the anyonic particle picture the counterclockwise circulation corresponds to the phase factor $e^{i2\pi s}$, where s is the effective spin of the anyon. Thus, a non-trivial spin

$$s = \frac{q\Phi}{2\pi} \tag{2.11}$$

is obtained whenever the charge, q, or the flux, Φ, are fractionalised. The values of the exchange statistics and the spin-phase factors emerging from this mechanical picture of anyons are consistent with the spin-statistics theorem (Pauli, 1940) that will be discussed in detail in Section 4.1.6.

As an extension, we can envisage the Aharonov–Bohm effect in terms of non-Abelian charges and fluxes. Non-Abelian charges can be described by vectors that span an n-dimensional Hilbert space. The vector components are also known as colour. Fluxes are then n-dimensional matrices that comprise a non-Abelian algebra. A circulation of a colour charge around a non-Abelian flux generates an n-dimensional unitary matrix U instead of merely a phase factor. Non-Abelian anyons can in general be decomposed into colour charges attached to localised non-Abelian fluxes. The circulation of one non-Abelian anyon around another can hence lead to a statistical evolution that non-trivially rotates the colour space of states. Such a description is given in Chapter 7.

Needless to say, non-Abelian gauge fields do not exist freely in nature. The corresponding charges and fluxes are encountered in the physics of quarks and gluons, which is known in high-energy physics as quantum chromodynamics. Their control is beyond our engineering capabilities. Nevertheless, they can emerge in carefully designed many-body systems. In general, the behaviour of topological systems can effectively be described by interacting charges and fluxes (Nussinov and Ortiz, 2009). Next we present how such gauge fields can emerge in the context of geometric phases.

2.2 Geometric phases and holonomies

The Aharonov–Bohm effect gives a consistent picture for the exotic statistical behaviour of Abelian anyons. Here we describe how Abelian and non-Abelian gauge fields can emerge in highly correlated quantum systems through the mechanism of geometric phases. Our understanding of geometric phases advanced significantly when Michael Berry presented an illuminating discussion in 1984 (Berry, 1984). In his seminal paper he showed that there are physical cases where the emerging non-dynamic phase factors cannot be eliminated by a gauge transformation. Hence, we are forced to assign a physical meaning to them.

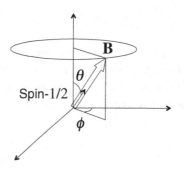

Fig. 2.3 A spin-1/2 particle is placed at a fixed position in the presence of a magnetic field $\mathbf{B}(\theta, \phi)$ that can take any arbitrary orientation. The spin of the particle adiabatically follows the orientation of the field.

2.2.1 Spin-1/2 particle in a magnetic field

2.2.1.1 The system

The emergence of effective gauge fields is best formulated in terms of the Abelian geometric or Berry phase (Berry, 1984; Pachos and Zanardi, 2001). Let us first present a simple example where a Berry phase appears that involves a static spin-1/2 particle in the presence of a magnetic field. Consider a magnetic field $\mathbf{B}(\theta, \phi)$ having orientation θ and ϕ, with $0 \leq \theta \leq \pi$ and $0 \leq \phi \leq 2\pi$, as shown in Figure 2.3, and constant non-zero magnitude B. A spin-1/2 particle is placed at the origin with spin orientation parallel to the field. Note that this situation is different from a charged particle in a magnetic field described above. Here, the magnetic field is introduced exclusively to control the orientation of the spin.

The Hamiltonian that describes the interaction between the magnetic field and the spin is given by

$$H = -\boldsymbol{\sigma} \cdot \mathbf{B}(\theta, \phi) = -\boldsymbol{\sigma} \cdot \hat{\mathbf{n}}(\theta, \phi)B, \tag{2.12}$$

where $\hat{\mathbf{n}}(\theta, \phi) = (\sin\theta\cos\phi, \sin\theta\sin\phi, \cos\theta)$ gives the orientation of the magnetic field and $\boldsymbol{\sigma} = (\sigma^x, \sigma^y, \sigma^z)$ are the Pauli matrices. An alternative way to write the Hamiltonian is

$$H = -\mathcal{U}(\theta, \phi)\sigma^z\mathcal{U}^\dagger(\theta, \phi)B = \mathcal{U}(\theta, \phi)H_0\mathcal{U}^\dagger(\theta, \phi), \tag{2.13}$$

where $H_0 = -B\sigma^z$ and

$$\mathcal{U}(\theta, \phi) = \begin{pmatrix} \cos\frac{\theta}{2} & e^{-i\phi}\sin\frac{\theta}{2} \\ e^{i\phi}\sin\frac{\theta}{2} & -\cos\frac{\theta}{2} \end{pmatrix} \tag{2.14}$$

is an SU(2) unitary rotation. The eigenstates of this system for any arbitrary orientation of the magnetic field are given by $|\uparrow(\theta, \phi)\rangle = \mathcal{U}(\theta, \phi)|\uparrow\rangle$ and $|\downarrow(\theta, \phi)\rangle = \mathcal{U}(\theta, \phi)|\downarrow\rangle$, with eigenvalues $E_\uparrow = B$ and $E_\downarrow = -B$ respectively, where $\sigma^z|\uparrow\rangle = |\uparrow\rangle$ and $\sigma^z|\downarrow\rangle = -|\downarrow\rangle$.

2.2.1.2 The geometric evolution

Consider the system being prepared initially in eigenstate $|\uparrow(\theta_0, \phi_0)\rangle$. When the orientation of the magnetic field changes slowly, the system adapts to the instantaneous eigenstate

$|\uparrow(\theta,\phi)\rangle$. This is guaranteed by adiabaticity that applies as long as the change in orientation of the magnetic field is slow compared to the characteristic energy scales of the system (Born and Fock, 1928). When the magnetic field comes back to its initial orientation (θ_0, ϕ_0), the Hamiltonian is again exactly the same as the initial one. So the state of the system is, up to an overall phase factor, equal to the initial state. The Schrödinger equation dictates that this phase factor is given by

$$e^{i\varphi} = e^{\oint_C \mathbf{A} \cdot d\mathbf{r}} e^{iE_\uparrow T}, \tag{2.15}$$

with

$$A_\mu = \langle \uparrow | \mathcal{U}^\dagger(\theta, \phi) \frac{\partial}{\partial \lambda^\mu} \mathcal{U}(\theta, \phi) | \uparrow \rangle. \tag{2.16}$$

The derivation of (2.15) and (2.16) is presented in the next subsection. The vector $\mathbf{A}$ is called Berry connection, or just connection, and it plays a similar role to the vector potential. Here $\mu = 1, 2$, $\lambda^\mu = \{\theta, \phi\}$, C is a cyclic path in this parametric space and T is the total time of the evolution. The first factor of the above phase, $\varphi_g = \frac{1}{i} \oint_C \mathbf{A} \cdot d\mathbf{r}$, is geometrical, in the sense that it depends only on the path spanned in the parametric space $\{\theta, \phi\}$. Hence, it is called geometric phase. Moreover, it is independent of the Hamiltonian H_0. The second term depends on the eigenvalue $E_\uparrow$ of the $|\uparrow(\theta, \phi)\rangle$ state and the total time of the evolution. We can remove it by subtracting an overall constant term from the Hamiltonian so that $E_\uparrow = 0$.

We can now explicitly evaluate the geometric phase resulting from the spin-1/2 particle in the magnetic field $\mathbf{B}(\theta, \phi)$. From (2.16) and (2.14) we have that the components of the connection $\mathbf{A}$ are given by

$$A_\theta = \langle \uparrow | \mathcal{U}^\dagger(\theta, \phi) \frac{\partial}{\partial \theta} \mathcal{U}(\theta, \phi) | \uparrow \rangle = \langle \uparrow | \begin{pmatrix} 0 & e^{-i\phi} \\ e^{i\phi} & 0 \end{pmatrix} | \uparrow \rangle = 0 \tag{2.17}$$

and

$$A_\phi = \langle \uparrow | \mathcal{U}^\dagger(\theta, \phi) \frac{\partial}{\partial \phi} \mathcal{U}(\theta, \phi) | \uparrow \rangle = \frac{i}{2} \langle \uparrow | \begin{pmatrix} 1 - \cos\theta & -\sin\theta e^{-i\phi} \\ -\sin\theta e^{i\phi} & -1 + \cos\theta \end{pmatrix} | \uparrow \rangle$$

$$= \frac{i}{2}(1 - \cos\theta). \tag{2.18}$$

The connection components corresponding to $|\downarrow\rangle$ can be evaluated analogously. They give the same values, but with an overall minus sign. The field strength, or curvature, corresponding to the connection $\mathbf{A}$ is therefore given by

$$F_{\theta\phi} = \partial_\theta A_\phi - \partial_\phi A_\theta = \frac{i}{2} \sin\theta, \tag{2.19}$$

with all other components being zero. Applying Stokes' theorem, as we did in (2.7), we obtain

$$\varphi_g = \frac{1}{i} \oint_C \mathbf{A} \cdot d\mathbf{r} = \frac{1}{i} \int \int_{S(C)} \mathbf{F} \cdot d\mathbf{s} = \frac{1}{2} \int \int_{S(C)} d\theta d\phi \sin\theta = \frac{1}{2}\Omega(C), \tag{2.20}$$

where C is the spanned loop in the parametric space $\{\theta, \phi\}$, $S(C)$ is the enclosed surface and $\Omega(C)$ is the solid angle of the loop spanned on the unit sphere by the vector $\hat{\mathbf{n}}(\theta, \phi)$, as shown in Figure 2.4.

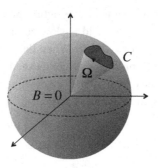

Fig. 2.4   In the case of a spin-1/2 particle in a magnetic field **B**, the loop C is spanned in the space of orientations $\{\theta, \phi\}$ of **B**. If the radius of the sphere parameterises the amplitude of the magnetic field then the point where the two spin states become degenerate is at the centre of the sphere where $B = 0$. The geometric phase is proportional to the solid angle, Ω, the loop C spans with respect to this point.

2.2.1.3 Properties of Berry's phase

The connection **A** and the quantum state $|\Psi\rangle$ can be subject to gauge transformations of the form

$$\mathbf{A}(\theta, \phi) \rightarrow \mathbf{A}(\theta, \phi) - \nabla \omega(\theta, \phi),$$
$$|\Psi(\theta, \phi)\rangle \rightarrow e^{i\omega(\theta, \phi)} |\Psi(\theta, \phi)\rangle, \tag{2.21}$$

where $\omega(\theta, \phi)$ is a scalar function. Nevertheless, Berry's phase, φ_{g}, is gauge-invariant. This follows from the application of Stokes' theorem (2.20) that relates it to the flux of the gauge-invariant field strength, **F**. The latter can take zero or non-zero values depending on whether **A** is a pure gauge or not.

The natural question arises: when can a system support non-vanishing **F**? To answer this we need to consider the geometry of the space of states $|\Psi\rangle$ parameterised by λ. From (2.20) we see that the geometric phase is proportional to a solid angle Ω. This solid angle is spanned by the loop C in the parametric space with respect to the singular point where the two states of the system become degenerate. For the case of a spin-1/2 particle consider the amplitude B of the magnetic field as one of the possible parameters, depicted as the radius of the sphere in Figure 2.4. Then $B = 0$, which corresponds to the centre of the sphere, makes $|\uparrow (\theta, \phi)\rangle$ and $|\downarrow (\theta, \phi)\rangle$ degenerate. Such singular points in the parametric space are necessary in order to create non-trivial curvature in the Hilbert space.

The process that gives rise to the geometric phase is equivalent, but not identical, to the mechanism that generates the phase of a charged particle when it is moving cyclically in the presence of a magnetic field. Similarly to a magnetic field, the geometric phase does not depend on the shape of the path as long as it circulates the same effective flux, as long as it is independent of the speed at which the path is traversed and as long as the adiabaticity condition is satisfied. The only difference is that here the cyclic evolution that gives rise to the geometric phase is not parameterised by the coordinates of the particle, but by some abstract parameters $\lambda^{\mu} = \{\theta, \phi\}$. Nevertheless, it is possible to create complex enough Hamiltonians so that this parametric space is formed by the coordinates of a quasiparticle. We shall consider such many-body Hamiltonians in later chapters. Now we want to see that the mechanism of geometric evolutions can give rise to more complex geometrical effects, such as non-Abelian phases.

2.2.2 Non-Abelian geometric phases

In this subsection we give the generalisation of Berry's phase (2.15) to the non-Abelian case. We also present how these phases can arise from the evolution operator corresponding to an adiabatic cyclic process.

2.2.2.1 The holonomy

To generalise geometric phases to the non-Abelian case we employ again unitary isospectral evolutions, as we did in the spin-1/2 case. Consider a parametric space $\mathcal{M} = \{\lambda^\mu, \mu = 1, \ldots, d\}$ and the D-dimensional Hamiltonian

$$H(\lambda(t)) = \mathcal{U}(\lambda(t)) H_0 \mathcal{U}(\lambda(t))^\dagger, \tag{2.22}$$

where $\mathcal{U}(\lambda) \in \mathrm{SU}(D)$ is a unitary rotation that changes in time. The λ's are classical parameters that can be controlled externally. We then assume that the ground state of H_0 consists of an n-dimensional degenerate subspace, $\mathcal{H}_0 = \{| \Psi^\alpha \rangle, \alpha = 1, \ldots, n\}$, with energy $E_0 = 0$ and an energy gap ΔE separating this subspace from the excited states.

To proceed we initiate the system in a certain state of $\mathcal{H}_0$. If one changes the parameters λ slowly in time compared to the energy gap ΔE, then the evolution is adiabatic. As above this implies that there is no transfer of population into the excited states. However, the transfer of population between the degenerate states of $\mathcal{H}_0$ is allowed. Indeed, the spanning of a loop C in the parametric space $\mathcal{M}$ results in an evolution that is an element of the n-dimensional unitary group $\mathrm{U}(n)$. This evolution acts on the ground state-space of the system in the following way:

$$| \Psi(C) \rangle = \Gamma_{\mathbf{A}}(C) | \Psi(0) \rangle, \tag{2.23}$$

where $| \Psi(0) \rangle$ and $| \Psi(C) \rangle$ both belong to $\mathcal{H}_0$. The non-Abelian geometric phase $\Gamma_{\mathbf{A}}(C)$, which is also known as a holonomy, is given by

$$\Gamma_{\mathbf{A}}(C) = \mathbf{P} \exp \oint_C \mathbf{A} \cdot d\lambda, \tag{2.24}$$

where $\mathbf{P}$ denotes path ordering. Here, and in the case of the geometric phase (2.15), the connection $\mathbf{A}$ is an anti-Hermitian operator as, for convenience, we absorbed in $\mathbf{A}$ an i factor. The matrix elements of its components are given by

$$(A_\mu)^{\alpha\beta} = \langle \Psi^\alpha | \mathcal{U}(\lambda)^\dagger \frac{\partial \mathcal{U}(\lambda)}{\partial \lambda^\mu} | \Psi^\beta \rangle. \tag{2.25}$$

This is the non-Abelian generalisation of the usual Berry phase (Berry, 1984) which was first discovered by Wilczek and Zee (1984).

2.2.2.2 Derivation of holonomy

We now present a derivation of (2.23), (2.24) and (2.25) that also provides a physical insight into the mechanism behind geometric evolutions. Consider the isospectral adiabatic evolution given in (2.22). Then the time evolution operator takes the form

$$U(0,T) = \mathbf{T} \exp\left(-i \int_0^T \mathcal{U}(\lambda) H_0 \mathcal{U}^\dagger(\lambda) dt \right). \tag{2.26}$$

Let us divide the time interval $[0, T]$ of the cyclic evolution into N equal segments Δt and define $\mathcal{U}_i = \mathcal{U}(\lambda(t_i))$ for $i = 1, \ldots, N$. As Δt becomes small in the limit of large N we have

$$U(0,T) = \mathbf{T} \lim_{N \to \infty} \exp\left(-i \sum_{i=1}^N \mathcal{U}_i H_0 \mathcal{U}_i^\dagger \Delta t \right) = \mathbf{T} \lim_{N \to \infty} \prod_{i=1}^N \mathcal{U}_i \exp\left(-i H_0 \Delta t \right) \mathcal{U}_i^\dagger. \tag{2.27}$$

In particular, the product $\mathcal{U}_i^\dagger \mathcal{U}_{i+1}$ of two successive unitaries gives rise to an infinitesimal rotation of the form

$$\mathcal{U}_i^\dagger \mathcal{U}_{i+1} \approx \mathbb{1} + \bar{\mathbf{A}}_i \cdot \Delta\boldsymbol{\lambda}_i, \tag{2.28}$$

where

$$(\bar{A}_i)_\mu = \mathcal{U}_i^\dagger \frac{\Delta \mathcal{U}_i}{\Delta(\lambda_i)_\mu} \tag{2.29}$$

with $\mu = 1, \ldots, d$. Hence, the evolution operator $U(0,T)$ becomes

$$U(0,T) = \mathbf{T} \lim_{N \to \infty} \mathcal{U}_N \left(\mathbb{1} - i H_0 N \Delta t + \sum_{i=1}^{N-1} \bar{\mathbf{A}}_i \cdot \Delta\boldsymbol{\lambda}_i \right) \mathcal{U}_1^\dagger. \tag{2.30}$$

The initial and the final transformations $\mathcal{U}_1$ and $\mathcal{U}_N$ are identical for the case of closed paths as they correspond to the same point of $\mathcal{M}$. With a reparameterisation they can be made equal to the identity transformation, i.e., $\mathcal{U}_1 = \mathcal{U}_N = \mathbb{1}$. We now consider the action of $U(0,T)$ on an initial state $|\Psi(0)\rangle$ that belongs to the ground state subspace, $\mathcal{H}_0$. If we demand adiabaticity, then at each time t_i the state $|\Psi(t_i)\rangle$ will have energy eigenvalue $E_0 = 0$. Then the action of H_0 from (2.30) is trivialised, thus obtaining

$$U(0,T) | \Psi(0)\rangle = \mathbf{T} \lim_{N \to \infty} \left(\mathbb{1} + \sum_{i=1}^{N-1} \mathbf{A}_i \cdot \Delta\boldsymbol{\lambda}_i \right) | \Psi(0)\rangle = \mathbf{P} \exp\left(\oint_C \mathbf{A} \cdot d\boldsymbol{\lambda} \right) | \Psi(0)\rangle. \tag{2.31}$$

In (2.31) the connection is defined by

$$\mathbf{A}(\lambda) = \Pi(\lambda)\bar{\mathbf{A}}(\lambda)\Pi(\lambda), \tag{2.32}$$

where $\Pi(\lambda)$ is the projector in the degenerate ground state subspace $\mathcal{H}_0(t)$ imposed by the adiabaticity condition. Hence, it is equivalent to (2.25). Notice that we replaced the time ordering $\mathbf{T}$ with the path ordering $\mathbf{P}$ as the parameter of the integration at the last expression is the position on the path C. This derivation makes it clear how the holonomy appears from the evolution operator by imposing the adiabaticity condition in a cyclic evolution. For the special case where the degenerate space is one-dimensional, $\dim(\mathcal{H}_0) = 1$, we obtain the Berry phase given in (2.15) and (2.16).

2.2.3 Properties of geometric evolutions

For completeness we now review some generic properties of holonomies (Pachos and Zanardi, 2001). For example, as the geometric evolutions resemble interactions with gauge fields, they inherit the property of gauge-invariance. Moreover, holonomies are parameterised by the loops C. Hence the properties of the loops, such as their composition, are reflected in the properties of the holonomies. Finally, the structure of the parametric space $\mathcal{M}$ of the holonomies determines the form of the connection $\mathbf{A}$.

2.2.3.1 Gauge transformations

For a Hamiltonian subject to isospectral transformations

$$H(\lambda) = \mathcal{U}(\lambda)H_0\mathcal{U}(\lambda)^\dagger, \tag{2.33}$$

a local gauge transformation is a unitary transformation

$$\mathcal{U}(\lambda) \to \mathcal{U}^g(\lambda) = \mathcal{U}(\lambda)g(\lambda) \tag{2.34}$$

with $g(\lambda) \in U(n)$, which does not change the form of Hamiltonian $H(\lambda)$. For non-trivial transformations this can only happen if $g(\lambda)$ is acting exclusively on the degenerate subspace $\mathcal{H}_0$. On this subspace the Hamiltonian acts trivially, i.e., it causes an evolution that is proportional to the identity. Hence, a gauge transformation $g(\lambda)$ should satisfy

$$g^{-1}(\lambda)H_0 g(\lambda) = H_0 \tag{2.35}$$

for all λ in $\mathcal{M}$. In terms of the unitary operator $\mathcal{U}(\lambda)$, the action of the gauge transformation is to merely reparameterise the λ variables. Taking into account that $gH_0 = H_0 g$ and $g\Pi = \Pi g$, we are able to obtain the gauge transformation of the connection as

$$\mathbf{A} \to \mathbf{A}^g = g^{-1}\mathbf{A}g + g^{-1}\mathbf{d}g. \tag{2.36}$$

It follows that the holonomy transforms as

$$\Gamma_\mathbf{A} \to \Gamma_{\mathbf{A}^g} = g^{-1}\Gamma_\mathbf{A}g. \tag{2.37}$$

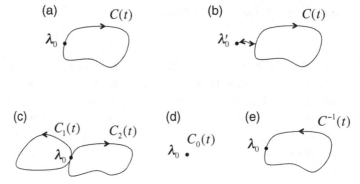

Fig. 2.5   (a) The loop $C(t)$ parameterised by $t \in [0, T]$ with base point λ_0. (b) The base point λ_0 of a loop can be moved to λ_0' by employing a path that is transversed forwards and backwards at the beginning and at the end of the loop, respectively. (c) The composite loop that spans C_1 for $t \in [0, T/2]$ and C_2 for $t \in [T/2, T]$. (d) The trivial loop C_0. (e) The inverse of a loop C is the same geometrical path, but it is spanned in the opposite direction.

Notice that in the new coordinates the state vectors $|\Psi\rangle$ become $|\Psi^g\rangle = g^{-1}|\Psi\rangle$. Thus, the action of the holonomic evolution on a state has an intrinsic (i.e., coordinate-free) meaning.

2.2.3.2 Loop parameterisation

Consider now the space $\mathcal{L}_{\lambda_0}$ of all loops based at a given point λ_0 of the parametric space $\mathcal{M}$, i.e., all loops C parameterised by $t \in [0, T]$ with $C(0) = C(T) = \lambda_0$, as shown in Figure 2.5(a). As long as the parametric space $\mathcal{M}$ is connected, the choice of λ_0 does not play any role, as it gives the same set of loops. Indeed, trivial loops (i.e., paths that are traversed forwards and backwards) can be employed to move the base point λ_0 to any position, as shown in Figure 2.5(b).

In the $\mathcal{L}_{\lambda_0}$ space we introduce a composition rule for loops

$$(C_2 \cdot C_1)(t) = \theta \left(\frac{1}{2} - \frac{t}{T} \right) C_1(2t) + \theta \left(\frac{t}{T} - \frac{1}{2} \right) C_2(2t - T), \qquad (2.38)$$

where $\theta(t)$ is the step function. In other words the composite loop runs successively through both constituent loops, as shown in Figure 2.5(c). The trivial loop element is identified with a point, i.e., $C_0(t) = \lambda_0$ for all $t \in [0, T]$, as shown in Figure 2.5(d). This is equivalent to a loop where a path is spanned both forwards and backwards without enclosing any area. As the inverse loop we denote $C^{-1}(t) = C(T - t)$, where $t \in [0, T]$, shown in Figure 2.5(e).

The basic properties of the holonomy $\Gamma_{\mathbf{A}}(C) = \mathbf{P} \exp \oint_C \mathbf{A} \cdot d\boldsymbol{\lambda}$ in terms of its loop dependence are the following. First, the composition of loops in $\mathcal{L}$ gives a holonomic evolution that is the product of the evolutions associated with the individual loops

$$\Gamma_{\mathbf{A}}(C_2 \cdot C_1) = \Gamma_{\mathbf{A}}(C_2)\, \Gamma_{\mathbf{A}}(C_1). \qquad (2.39)$$

Second, staying at the same point at all times corresponds to the trivial holonomy

$$\Gamma_{\mathbf{A}}(C_0) = \mathbb{1}. \qquad (2.40)$$

Third, traversing the path C in reverse order gives the inverse holonomy

$$\Gamma_{\mathbf{A}}(C^{-1}) = \Gamma_{\mathbf{A}}^{-1}(C). \tag{2.41}$$

Fourth, changing the functional dependence of the path in terms of t does not change the value of the holonomy as long as the adiabaticity condition holds

$$\Gamma_{\mathbf{A}}(C \circ f) = \Gamma_{\mathbf{A}}(C), \tag{2.42}$$

where f is any function of t. This means that the rate with which the path C is spanned does not change the value of $\Gamma_{\mathbf{A}}(C)$. Hence, the group properties of the holonomies follow simply from the geometric properties of the spanned loops. This equivalence allows us to efficiently represent holonomies by their corresponding loops.

2.2.3.3 Holonomies as unitary matrices

The path dependence of the holonomies demonstrates that the set of holonomies spanned by all paths (i.e., $\mathrm{Hol}(\mathbf{A}) = \Gamma_{\mathbf{A}}(\mathcal{L})$) is a subgroup of $\mathrm{U}(n)$. Such a subgroup is known as the holonomy group of the connection $\mathbf{A}$. When the holonomy group coincides with the whole $\mathrm{U}(n)$, then the connection $\mathbf{A}$ is called irreducible. In order to determine if this is the case for a given connection it is useful to consider the curvature, or effective magnetic field, $\mathbf{F}$, of the connection $\mathbf{A}$ with components

$$F_{\mu\nu} = \partial_\mu A_\nu - \partial_\nu A_\mu + [A_\mu, A_\nu]. \tag{2.43}$$

The relation between the curvature and the irreducibility of the connection is given by the following statement (Nakahara, 2003): all possible $F_{\mu\nu}$'s can be considered as the basis algebra elements that generate the holonomic unitary group.

It follows in particular that when the $F_{\mu\nu}$'s span the whole $\mathrm{U}(n)$ algebra, then the connection is irreducible. The possibility of spanning the whole group $\mathrm{U}(n)$ by a non-Abelian geometric phase is related to quantum computational universality. Indeed, if holonomies are employed to perform quantum computation then we need to know if spanning any arbitrary sequence of paths in $\mathcal{L}$ is sufficient to generate any desired quantum algorithm. Irreducibility of the connection provides a mathematical criterion to answer this question.

Finally, let us present some structural properties of the holonomies. The holonomy, $\Gamma_{\mathbf{A}}(\mathcal{L})$, is exclusively acting on the degenerate states that are actually evolved by the adiabatic process. This is apparent from (2.25), which projects $\mathcal{U}^\dagger \partial \mathcal{U}/\partial \lambda^\mu$ on the degenerate subspace. If, during the spanning of the whole path, certain states are not involved in the evolution then the resulting holonomy $\Gamma_{\mathbf{A}}(\mathcal{L})$ will act trivially on them. In other words, if $\mathcal{U}$ acts as an identity on a certain state of the degenerate subspace, then the holonomy will also act trivially on it.

Furthermore, the holonomy operator arises from the non-commutativity of the control transformations which produce effectively a curvature. Consider as an example the case of the geometric phase produced in front of the spin states of an electron in the presence of a rotated magnetic field. The non-commutativity here is between the different $\mathrm{U}(2)$ control unitaries given by (2.14). These unitary matrices change the orientation of the magnetic field parameterised by θ and ϕ, as shown in Figure 2.3. This characteristic also holds in the non-Abelian generalisation of the geometric phase.

Finally, note that in the absence of adiabaticity the connection **A** is not projected to any subspace of states of the Hamiltonian. Thus it is a pure gauge that gives a trivial holonomy. Only when it is non-trivially projected in a certain subspace of states does it give rise to a non-trivial geometric phase.

2.2.4 Anyons and geometric phases

The description of anyons in terms of the Aharonov–Bohm effect has conceptual value, but it can offer little to the physical realisations of anyons. On the other hand, geometric phases provide an efficient mechanism for anyonic statistics that can be met in physically plausible topological systems. We now present the main characteristics of topological systems that allow their description in terms of geometric phases. These characteristics are illustrated later with concrete examples.

As we have seen, topological systems are two-dimensional many-body systems with extended wave functions and localised quasiparticle excitations. To describe statistical evolutions in terms of geometric phases we need to identify the control parameter space $\mathcal{M}$ with the coordinates of the quasiparticle, x^μ. Moreover, the effective magnetic field, **F**, arising from the geometric connection, **A**, should be tightly confined at the position of the quasiparticles. Finally, the generation of a phase factor due to the flux of **F** implies a charge degree of freedom. Consider now braiding evolutions between quasiparticles. These evolutions correspond to loops in the coordinate parametric space $\mathcal{M}$. Hence, their statistical evolution can be described by a geometric phase. The generated phases are independent of the shape of the path due to the confinement of the effective flux at the quasiparticle position. Their description can therefore be given by Figure 2.2(a), where the field **F** gives rise to the flux Φ. This approach facilitates the description of anyons in terms of the Aharonov–Bohm effect through the language of geometric phases.

But how can we engineer a system that exhibits a geometric phase with confined effective flux? In the above example of a spin-1/2 particle in the magnetic field, the connection **A** given in (2.18) has a sinusoidal dependence on the control parameters $\{\theta, \phi\}$. This is a rather general property. A finite system is isospectrally transformed by unitary matrices that are parameterised by periodic, e.g., trigonometric functions. This results from the compactness of the parametric space $\mathcal{M}$ and the finite dimensionality of the Hilbert space. Hence, the resulting effective magnetic field **F** has a similar functional dependence. To acquire a strictly confined effective field we need a non-compact parametric space. Intuitively, this requires an infinitely large Hilbert space, e.g., a system that comprises an infinite number of constituent particles. To be able to go beyond the sinusoidal dependence the isospectral unitary rotations need to act non-trivially on infinitely many states of the system resulting, in general, in highly correlated states. In practice, very large but finite systems are expected to exhibit exponentially confined effective flux.

Beyond the paradigm presented here there exist topological systems with statistical evolutions that do not require the adiabatic condition. In the following chapters we see anyonic systems, where the parametric space is discrete and the adiabatic condition does not, strictly speaking, apply. Moreover, we also present examples where the statistical evolution can be calculated via geometric phases. The first explicit description of anyonic statistics in terms of geometric phases was given by Arovas *et al.* (1984). They showed how the statistics of Abelian anyons, appearing in the fractional quantum Hall effect, can be expressed in terms of a Berry phase. How to derive the non-Abelian statistics from geometric phases is an active topic of research (Bolukbasi and Vala, 2011; Chung and Stone, 2007; Lahtinen and Pachos, 2009; Read, 2009; Tserkovnyak and Simon, 2003).

2.3 Example I: Integer quantum Hall effect

We now present the quantum mechanical description of a charged particle confined to a plane in the presence of a magnetic field. This simple system has a surprisingly rich behaviour (Stern, 2008). What emerges is a new phenomenon known as the integer quantum Hall effect. In 1985, Klaus von Klitzing was awarded the Nobel prize for its experimental discovery (von Klitzing *et al.*, 1980). He noticed that the Hall conductivity has a step-like behaviour, with distinct plateaus, as a function of the inverse magnetic field strength (Hastings and Michalakis, 2009). This is unlike the linear behaviour that one expects to find classically. A unique insight into the physics of the integer quantum Hall effect is given in terms of geometric phases through Laughlin's thought experiment (Laughlin, 1981).

2.3.1 Wave function of a charged particle in a magnetic field

Consider a spinless electron with charge e and mass m_e confined in a thin film. The film has dimensions L_1, L_2 and L_3 and is positioned on the x–y plane, as shown in Figure 2.6. Suppose that the electrons are in a coherent quantum state throughout the sample and a

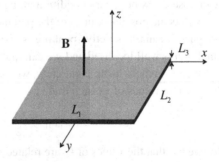

Fig. 2.6 The magnetic field **B** is aligned along the z direction, while the electron is restricted to move in a thin film with dimensions L_1, L_2 and L_3 that lies on the x–y plane.

constant magnetic field along the z direction $\mathbf{B} = (0, 0, B)$ is applied. In the Landau gauge the vector potential is then given by $\mathbf{A} = (0, Bx, 0)$ and the Schrödinger equation (2.4) becomes

$$-\frac{\hbar^2}{2m_e}\left[\frac{\partial^2}{\partial x^2} + \left(\frac{\partial}{\partial y} + i\frac{eB}{\hbar}x\right)^2 + \frac{\partial^2}{\partial z^2}\right]\Psi = E\Psi. \tag{2.44}$$

If the electron is subject to periodic boundary conditions along the y and z directions it is convenient to write the wave function in the form $\Psi(x, y, z) = \Phi(x)e^{ik_y y}e^{ik_z z}$ with $k_y = p2\pi/L_2$, $k_z = q2\pi/L_3$, where p and q are integers. Then the Schrödinger equation (2.44) reduces to

$$-\frac{\hbar^2}{2m_e}\frac{\partial^2\Phi(x)}{\partial x^2} + \frac{\hbar^2}{2m_e}\left(k_y + \frac{eB}{\hbar}x\right)^2\Phi(x) = E'\Phi(x), \tag{2.45}$$

where $E' = E - \hbar^2 k_z^2/(2m_e)$. This expression is equivalent to the Schrödinger equation of a one-dimensional harmonic oscillator with its origin shifted in position by

$$x_0 = -\frac{\hbar k_y}{eB} \tag{2.46}$$

and an effective spring constant given by $\gamma_{\text{eff}} = e^2 B^2/m_e$. Its frequency $\omega = \sqrt{\gamma_{\text{eff}}/m_e}$ is precisely the cyclotron frequency of the classical circular orbit of an electron in a constant magnetic field, $\omega = eB/m_e$ (see Exercise 2.2). The solution of the initial Schrödinger equation is hence given by

$$\Psi(x, y, z) = \Psi_n^{\text{HO}}(x - x_0)\sqrt{\frac{1}{L_2}}e^{ik_y y}\sqrt{\frac{1}{L_3}}e^{ik_z z}, \tag{2.47}$$

where $\Psi_n^{\text{HO}}(x - x_0)$ is the nth eigenstate of the harmonic oscillator and the energy levels are given by

$$E = (n + \frac{1}{2})\hbar\omega + \frac{\hbar^2}{2m_e}k_z^2, \tag{2.48}$$

with $n = 0, 1, 2, \ldots$ The states that correspond to different values of n are called Landau levels (Landau and Lifshitz, 1977). For each n, these states are parameterised by the discrete values of k_y and k_z.

We now impose a box boundary condition along the x direction given by $-L_1/2 < x < L_1/2$. This restricts the possible values of the parameter x_0. Since $\Psi_n^{\text{HO}}(x - x_0)$ is extended throughout space it cannot, strictly speaking, satisfy the box boundary conditions. However, for small n it is well localised and we can make the crude approximation of ignoring the effect of this mismatch. More concretely, we assume that the boundary condition just restricts the possible values of x_0 such that

$$|x_0| < \frac{L_1}{2}. \tag{2.49}$$

From (2.46) we see that the values of k_y are related to x_0. So they will also be restricted by

$$|k_y| < \frac{eBL_1}{2\hbar}. \tag{2.50}$$

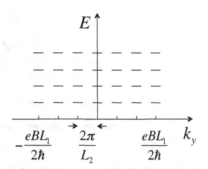

Fig. 2.7   The energy structure of a particle confined in a thin film in the presence of a perpendicular magnetic field. There is a large degeneracy at each energy level, called a Landau level, parameterised by the discrete k_y. The states are depicted by lines, rather than points, for illustration purposes.

As shown in Figure 2.7, the spacing between the values of k_y is $2\pi/L_2$ due to the finite size of the sample. So there is a finite number, N_{k_y}, of possible states with different values of k_y that can be accommodated along the x direction. Note that all of these states have exactly the same energy as E is independent of k_y. Indeed, the number of possible k_y values associated with each Landau level is

$$N_{k_y} = 2 \frac{eB}{\hbar} \frac{L_1}{2} \frac{1}{\frac{2\pi}{L_2}} = \frac{L_1 L_2}{2\pi \frac{\hbar}{eB}} = \frac{\text{Area of sample}}{2\pi l_B^2}, \tag{2.51}$$

where

$$l_B = \sqrt{\frac{\hbar}{eB}} \tag{2.52}$$

is the magnetic length. The energy spacing between Landau levels is

$$\Delta E = \hbar\omega = \frac{eB\hbar}{m_e} = \frac{\hbar^2}{m_e l_B^2}. \tag{2.53}$$

Hence, an electron confined to move in two dimensions and in the presence of a perpendicular magnetic field has discrete energy levels and a large degeneracy at each one of them, as illustrated in Figure 2.7.

2.3.2   Current behaviour and Hall conductivity

We now describe the integer quantum Hall effect (von Klitzing *et al.*, 1980) based on the previous solution of the Schrödinger equation and some rather classical analysis. Consider the configuration of Figure 2.8, where a metallic sample is placed on the x–y plane with a magnetic field, $\mathbf{B} = (0, 0, B)$, along the z direction and an electric field $\mathbf{E} = (0, E_y, 0)$ giving rise to a potential difference along the y direction. The sample provides a two-dimensional electron gas that is kept at low temperature. The conductivity dictates that the produced

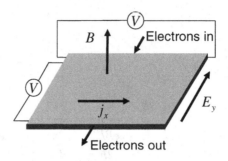

Fig. 2.8 The experimental configuration that gives rise to the quantum Hall effect. A magnetic field is applied perpendicularly to the two-dimensional sample, in which electrons are free to move. A potential difference along the y direction is created by an electric field and electrons are free to enter and exit the sample. The currents along the x direction, j_x, and the y direction, j_y, are measured by voltmeters.

current density due to the electric field is given by $\mathbf{j} = \sigma_0\mathbf{E}$. As the electric field is only along the y direction, we have the non-zero current component

$$j_y = \sigma_0 E_y. \tag{2.54}$$

Here the conductivity constant is given by

$$\sigma_0 = \frac{n_e e^2 \tau_0}{m_e^*}, \tag{2.55}$$

where n_e is the electron density, τ_0 corresponds to an average time between collisions of the electron with imperfections in the sample and m_e^* is the effective mass of the electron in the material (Gasiorowicz, 1996). In the presence of the magnetic field, the electrons experience a Lorentz force

$$\mathbf{F}_L = -e\mathbf{v} \times \mathbf{B}. \tag{2.56}$$

As $\mathbf{j} = -en_e\mathbf{v}$, the extra force due to the magnetic field can be written as $\mathbf{F}_L = \mathbf{j} \times \mathbf{B}/n_e$. Alternatively, one can view the Lorentz force as produced by an equivalent electric field $\mathbf{E}' = -\mathbf{F}_L/e$. Including the effect of the magnetic field, we obtain the following current density:

$$\mathbf{j} = \sigma_0 \left(\mathbf{E} - \frac{\mathbf{j} \times \mathbf{B}}{n_e e}\right). \tag{2.57}$$

Finally, the density of electrons can be given in terms of the maximum number of electrons per unit volume that can be accommodated in the Landau levels. From (2.51) we have

$$n_e = f\frac{N_{k_y}}{L_1 L_2} = f\frac{eB}{h}, \tag{2.58}$$

where f is a proportionality constant to be determined. Substituting back the components of the current we finally obtain

$$\sigma_{yy} = \frac{j_y}{E_y} = \frac{n_e e^2 \tau_0/m_e^*}{1 + (e\tau_0 B/m_e^*)^2} \quad , \quad \sigma_{xy} = \frac{j_x}{E_y} = f\frac{e^2}{h}\left(1 - \frac{1}{1 + (e\tau_0 B/m_e^*)^2}\right), \tag{2.59}$$

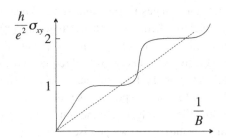

Fig. 2.9 An illustration of the measured Hall conductivity σ_{xy} (solid line) in units of e^2/h as a function of $1/B$. Plateaus emerge that correspond to the integer values of $f = \sigma_{xy}h/e^2$. Classically, a linear behaviour is expected (dotted line), so the plateaus are a purely quantum mechanical effect.

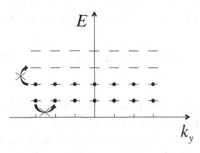

Fig. 2.10 Completely filled Landau levels up to a certain energy, where electrons are denoted by black dots. Elastic collisions of the electrons with impurities cannot cause the electrons to move to another filled state within the same Landau level due to Pauli's exclusion principle. Transitions to higher Landau levels are energetically forbidden due to low temperature.

where we defined as σ_{yy} the longitudinal and as σ_{xy} the transverse conductivity. In the absence of B we recover $\sigma_{yy} = \sigma_0$ and $\sigma_{xy} = 0$ as in (2.54). This shows that the magnetic field changes the direction of the current. It was found experimentally by von Klitzing *et al.* (1980) that for varying B the parameter f takes only integer values and the conductivities become

$$\sigma_{yy} = 0, \quad \sigma_{xy} = f\frac{e^2}{h}. \tag{2.60}$$

This deviates from the classically expected behaviour of the Hall conductivity, which is given by (2.59), and is illustrated in Figure 2.9. Below we explain why the conductivity σ_{yy} becomes zero. In the next subsection we present Laughlin's thought experiment that explains why σ_{xy} takes quantised values.

The unexpected behaviour of the conductivities σ_{yy} and σ_{xy} can be interpreted in the following simplified way. Consider a metallic sample, with the quantum structure of the Landau levels and the large degeneracy given by (2.51). We assume that the sample has many electrons and the system is prepared into its ground state and kept at very low temperature. As the electrons are fermions, each one can occupy only one quantum state, thus filling up one by one the Landau levels, as shown in Figure 2.10. When an electron is moving in the sample it has an average time between collisions with impurities of the lattice given by τ_0. Such a collision causes the electron to scatter elastically. During this process the electron changes its quantum state. When a certain number of Landau levels are already completely filled, an electron cannot undergo such a scattering with an impurity, as all states with the same energy are completely filled. Moving an electron to an empty Landau level with higher energy would require extra energy that needs to be obtained from outside the system in the form of thermal fluctuations. When we keep the temperature of the sample very low, such a process is prohibited. So, the electron is not allowed to collide with impurities. Thus, effectively the time between collisions becomes infinite, $\tau_0 \to \infty$. Substituting this into (2.59) we precisely obtain (2.60), which is the desired result.

Equation (2.60) suggests how to measure the ratio e^2/h experimentally. It has been shown that the integer quantum Hall effect provides the means to measure this ratio with the impressive accuracy of one part in 10^7. This has applications to metrology, for the calibration of resistance in units of e^2/h. Moreover, it determines independently the fine structure constant $\alpha = e^2/(2hc\epsilon_0)$ of quantum electrodynamics (NIST, 2006).

2.3.3 Laughlin's thought experiment and geometric phases

To theoretically understand the quantal nature of the conductivity σ_{xy}, we resort to Laughlin's thought experiment (Laughlin, 1981). This experiment gives a new perspective onto the quantum Hall effect by connecting it to geometric phases. The possibility of explaining the quantisation of σ_{xy} with general principles is based on the following observation. In the previous analysis of the quantum Hall system we argued that impurities play a central role in the emergence of conductivity plateaus. But impurities appear in a random way throughout the sample. So we cannot expect a microscopic analysis of the sample to explain why all possible values of the conductivity σ_{xy} are quantised. This suggests that there should be a macroscopic principle governing the behaviour of the system.

Let us return to the initial configuration of the quantum Hall system. We shall be interested in samples with cylindrical configurations, such as the one shown in Figure 2.11. Assume that we can continuously change the boundary conditions of the sample from rectangular, given in Figure 2.12(a), to cylindrical, given in Figure 2.12(b) or (c). If the system is large enough and the energy gap above the ground state remains finite at all times, then the generic properties of the system, such as its conductivity, are not expected to change

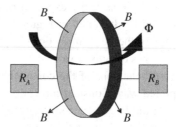

Fig. 2.11   Configuration of Laughlin's thought experiment. Increasing adiabatically the flux Φ by a single flux quantum, Φ_0, brings the looping sample back to its original quantum state, up to a geometric phase. This phase has a topological behaviour that explains why the quantum Hall conductivity takes discrete values. During the flux increase, charge is transported between the reservoirs R_A and R_B due to the quantum Hall effect.

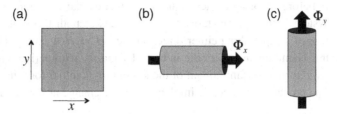

(a)   (b)   (c)

Fig. 2.12 (a) A two-dimensional sample of rectangular shape used for the quantum Hall effect. (b) We slowly deform the rectangle to a cylindrical shape oriented along the x direction, while we adiabatically turn on a flux Φ_x. (c) The rectangle can alternatively be deformed to a cylinder along the y direction with a flux Φ_y threading through it.

when we continuously change the boundary conditions (Niu and Thouless, 1987). Consider a magnetic flux Φ that is not in contact with the rectangular sample. A change in this flux corresponds to a gauge transformation as the magnetic field never comes in contact with the sample. Consider now the cylindrical configuration with the flux Φ threading the cylinder, as shown in Figure 2.11. A change in the flux Φ induces a current along the ribbon due to Faraday's law. Still, when the value of the flux is increased by integer multiples of the unit flux

$$\Phi_0 = \frac{hc}{e}, \tag{2.61}$$

then it corresponds to a gauge transformation, as deduced from the discussion below equation (2.7).

Due to Faraday's law (see Exercise 2.3), the change in the magnetic flux Φ induces an electric field along the ribbon. Together with the magnetic field **B**, which is perpendicular to the sample, they give rise to the quantum Hall effect. So while the flux is changing we expect a current to emerge between the reservoirs R_A and R_B of Figure 2.11. An increase by a unit flux Φ_0 gives rise to the transfer of charge

$$Q = \sigma_{xy}\frac{h}{e}. \tag{2.62}$$

Here we define the Hall conductivity to be given by $\sigma_{xy} = fe^2/h$, with f being an unknown parameter. Classically, f is an integer as at each cycle the charge comes in units of the electronic charge, e. Translated to the quantum world this argument implies that the conductivity is related to the average transported charge that need not be an integer multiplet of e. This would be the case if, for example, different flux cycles could lead to different integer charge transfers, resulting in a non-integral expectation value. We would like to show that the conductivity and, as an extension, the average transferred charge comes indeed in quantised steps.

To investigate the behaviour of the average conductivity we need to evaluate the expectation value of the current operator J. To do so we consider the adiabatic evolution where we change the boundary condition of the system by deforming the rectangular sample, shown in Figure 2.12(a), to a cylinder oriented along the x direction, as shown in Figure 2.12(b). We then introduce a threading flux Φ_x through the cylinder, before we bring the system back to its original rectangular configuration. Then we deform the system to a cylinder oriented along the y direction, as shown in Figure 2.12(c), introduce a flux Φ_y through it, before we bring it back again to the rectangular configuration. The Hamiltonian of the system is then parameterised by the values of both fluxes Φ_x and Φ_y that threaded the looped sample along either direction, i.e., $H = H(\Phi_x, \Phi_y)$. It is important to notice that this Hamiltonian is periodic in both Φ_x and Φ_y with period Φ_0, as increasing the flux by Φ_0 returns the Hamiltonian of the system to its initial form up to a gauge transformation. The process of physically implementing the change in flux induces the non-zero currents

$$J_x = c\frac{\partial H}{\partial \Phi_x} \text{ and } J_y = c\frac{\partial H}{\partial \Phi_y}. \tag{2.63}$$

Let us assume that the system is prepared initially at its ground state $\big| \Psi(\Phi_x, \Phi_y) \big\rangle$ whose energy, E_0, does not change when we vary Φ_x or Φ_y. Moreover, we demand that the ground state of the system is separated from the excited states by a non-zero energy gap at all times. Then, adiabatic evolutions can be realised by varying the variables slowly with respect to this gap. In particular, we consider the evolution where we first increase the value of Φ_x by Φ_0 and then the value of Φ_y by Φ_0. These flux increases are accompanied by the corresponding cylindrical deformations. Due to gauge-invariance the final state of the electrons should be equal to the initial one up to an overall geometric phase, i.e.,

$$\big| \Psi(\Phi_x + \Phi_0, \Phi_y + \Phi_0) \big\rangle = e^{i\varphi_g} \big| \Psi(\Phi_x, \Phi_y) \big\rangle. \tag{2.64}$$

To evaluate the geometric phase φ_g we employ the following methodology. The Schrödinger equation for the evolution when only Φ_x is varied becomes

$$i\hbar\partial_t | \Psi \rangle = i\hbar\partial_t\Phi_x \big| \partial_{\Phi_x}\Psi \big\rangle = H | \Psi \rangle. \tag{2.65}$$

This gives

$$i\hbar\big\langle \partial_{\Phi_y}\Psi \big| \partial_t\Phi_x \big| \partial_{\Phi_x}\Psi \big\rangle = \big\langle \partial_{\Phi_y}\Psi \big| H | \Psi \rangle, \tag{2.66}$$

where $| \Psi \rangle = \big| \Psi(\Phi_x, \Phi_y) \big\rangle$ and $\partial_a = \partial/\partial a$. Taking the complex conjugate of (2.66) and adding it to itself, we obtain

$$-\hbar\partial_t\Phi_x 2\mathrm{Im}(\langle \partial_{\Phi_y}\Psi | \partial_{\Phi_x}\Psi \rangle) = \big\langle \partial_{\Phi_y}\Psi \big| H | \Psi \rangle + \langle \Psi | H \big| \partial_{\Phi_y}\Psi \big\rangle. \tag{2.67}$$

As H is a Hermitian operator, we have

$$\partial_{\Phi_y}(\langle \Psi | H | \Psi \rangle) = \big\langle \partial_{\Phi_y}\Psi \big| H | \Psi \rangle + \langle \Psi | \partial_{\Phi_y}H | \Psi \rangle + \langle \Psi | H \big| \partial_{\Phi_y}\Psi \big\rangle. \tag{2.68}$$

We now use $\partial_{\Phi_y}(\langle \Psi | H | \Psi \rangle) = \partial_{\Phi_y}(E_0\langle \Psi | \Psi \rangle) = \partial_{\Phi_y}E_0 = 0$, which applies due to the assumption that the ground state energy, E_0, is independent of the flux Φ_y. Substituting this into (2.67) and employing (2.63), we finally obtain

$$\langle \Psi | J_y | \Psi \rangle = c\hbar\partial_t\Phi_x K(\Phi_x, \Phi_y), \tag{2.69}$$

where

$$K(\Phi_x, \Phi_y) = 2\mathrm{Im}(\langle \partial_{\Phi_y}\Psi | \partial_{\Phi_x}\Psi \rangle). \tag{2.70}$$

Notice that $\partial_t \Phi_x / c$ is the electromotive force that we can relate through the Schrödinger equation to the Hall current $\langle \Psi | J_y | \Psi \rangle$ with the proportionality factor $\hbar c^2 K$ (see Exercise 2.3). This factor is the transverse Hall conductivity. Hence

$$\sigma_{xy}(\Phi_x, \Phi_y) = \hbar c^2 K(\Phi_x, \Phi_y). \tag{2.71}$$

If we introduce the connection corresponding to the variations of the ground state with respect to the parameter $\{\Phi_x, \Phi_y\}$ given by

$$A_{\Phi_x} = \langle \Psi | \partial_{\Phi_x} \Psi \rangle \text{ and } A_{\Phi_y} = \langle \Psi | \partial_{\Phi_y} \Psi \rangle, \tag{2.72}$$

then the non-zero component of the field strength is

$$F_{\Phi_x \Phi_y} = \partial_{\Phi_x} A_{\Phi_y} - \partial_{\Phi_y} A_{\Phi_x} = 2i \text{Im}(\langle \partial_{\Phi_x} \Psi | \partial_{\Phi_y} \Psi \rangle). \tag{2.73}$$

Hence, the Hall conductivity can be expressed in terms of the curvature of the system's Hilbert space parameterised by the fluxes $\{\Phi_x, \Phi_y\}$, i.e.,

$$K(\Phi_x, \Phi_y) = -i F_{\Phi_x \Phi_y}. \tag{2.74}$$

Equation (2.20) relates the geometric phase φ_g to the field strength $\mathbf{F}$. Hence, the total conductivity is proportional to the geometric phase resulting from the adiabatic evolution of increasing both fluxes Φ_x and Φ_y by Φ_0.

To evaluate the total transfer of charge we can calculate the total conductivity through an adiabatic transport (see Exercise 2.3). For that we need to integrate the Φ_x and Φ_y variables over their whole range $\Sigma = [0, \Phi_0] \times [0, \Phi_0]$. Hence, the total conductivity is proportional to

$$\nu = \frac{1}{2\pi} \int \int_\Sigma K(\Phi_x, \Phi_y) d\Phi_x d\Phi_y. \tag{2.75}$$

Note that due to periodicity, Σ corresponds to a torus. We now employ the Gauss–Bonnet formula (Chern, 1944) that states

$$\frac{1}{2\pi} \int \int_S \mathcal{K} ds = 2 - 2g, \tag{2.76}$$

where S is a general compact surface with surface element ds, $\mathcal{K}$ is its local curvature and g is the number of its handles, also known as the genus of the surface. As it happens (Avron *et al.*, 2003) this equation also holds if we use the Hilbert space curvature $K(\Phi_x, \Phi_y)$ in place of the geometric curvature $\mathcal{K}$. In this case the right-hand side of (2.76) is an integer, but it does not necessarily relate to the genus of a two-dimensional surface. Hence, the Hall conductivity comes in quantised steps and the average charge (2.62) evaluated quantum mechanically is also quantised. The integer ν is known as the Chern number (Bohm *et al.*, 2003).

Summary

In this chapter we presented the Aharonov–Bohm effect that describes the quantum mechanical evolution of charged particles in the presence of magnetic fields. We also considered geometric phases that emerge when a quantum system evolves in a cyclic adiabatic fashion. They are described in terms of effective magnetic fields that can actually be Abelian or non-Abelian. Geometric phases are also known as Berry phases or holonomies.

We argued that the anyonic statistics of many topological systems can be determined in terms of geometric phases. Such an approach has already proven fruitful for understanding the nature of anyonic statistics as well as giving the means to carry out quantitative studies. Explicit demonstration of the anyonic statistics is the ultimate criterion for proving the topological character of a model. In Chapter 6 we employ geometric phases to evaluate the non-Abelian statistics of such a concrete model.

We have seen that to support geometric phases with topological properties, a system needs to comprise a large number of constituent particles. In this case the evolutions are tolerant to erroneous deformations of the traversed path. This is a topological characteristic in the behaviour of the evolution. The underlying redundancy in the encoding and this resilience to deformation errors resembles quantum error correction. In the latter, quantum information is encoded in a redundant way so that errors can be neutralised. The larger the quantum system the more protected the information becomes. The analogy between topological systems and quantum error correction is made explicit in Chapter 5.

Exercises

2.1 Consider a quantum system with three states in Λ-configuration subject to the Hamiltonian

$$H = \begin{pmatrix} \Delta & \Omega_1 & \Omega_2 \\ \Omega_1^* & 0 & 0 \\ \Omega_2^* & 0 & 0 \end{pmatrix}. \tag{2.77}$$

Here Ω_1 and Ω_2 are complex numbers that correspond to the coupling between the states and Δ is a real energy shift. Evaluate all the possible holonomies that can be produced by manipulating Ω_1 and Ω_2. [*Hint*: Determine first the degeneracy conditions of the Hamiltonian.]

2.2 Consider an electron with charge e and mass m_e confined to move on the plane in the presence of a perpendicular magnetic field B. Employ the Lorentz force (2.56) to show that the classical motion of the electron is circular with frequency

$$\omega = \frac{eB}{m_e}, \tag{2.78}$$

which is known as the cyclotron frequency. Demonstrate the correspondence principle between this classical description and the quantum one given in (2.47). [*Hint*: For the last part compare the evolutions with large radius for the classical case with the quantum states that give large mean radius.]

2.3   Consider the cylindrical sample given in Figure 2.11 with radius R. Employ Faraday's law of induction

$$\oint \mathbf{E} \cdot d\mathbf{r} = -\frac{1}{c}\frac{\partial \Phi}{\partial t} \tag{2.79}$$

to determine the value of the electric field along the ribbon generated by the change in the magnetic flux. Due to the quantum Hall effect a current is expected to emerge between the two reservoirs R_A and R_B with density

$$j(t) = \sigma_{xy}E(t). \tag{2.80}$$

By integrating the current contributions along the whole ribbon show that the total current that flows between the reservoirs is given by

$$J(t) = \frac{\sigma_{xy}}{c}\frac{\partial \Phi(t)}{\partial t}. \tag{2.81}$$

Finally, show that the total transfer of charge, when the flux is increased by $\Delta\Phi = \Phi_0 = hc/e$, is

$$Q = \sigma_{xy}\frac{h}{e}. \tag{2.82}$$

3

Quantum computation

Computation is a process of performing a large number of simple operations. The main building blocks of classical computers are bits. Each of them can take either the value 0 or 1. A series of gates can change these values by acting on them selectively. This process is structured to fulfil a purpose. For example, if we are interested in adding two numbers, the numbers are encoded into bits and gates are performed such that the final state of the bits reveals the desired answer: a number which is the sum of the initial ones. While it is possible to build algorithms that can perform almost any computation, it is also desirable to find the answer within a reasonable length of time or with a reasonable amount of resources. There is an ever-growing demand in computational power for both scientific and commercial purposes. Indeed, it is surprisingly easy to find an application that can jam even the fastest computer. This fuels a vast effort in the research of information science to increase the speed and processing power of computers.

Modern computational models are based on the universal Turing machine (Turing, 1937). This is a theoretical information processing model that employs the elementary gate processes described above. It can efficiently simulate any other device capable of performing an algorithmic process. Since the introduction of the Turing machine, physics has influenced computation in many ways. Deterministic or probabilistic computation, the cloning principle and irreversibility are well-known principles of the classical world that almost subliminally passed into the structures of programs (Bennett 1982). The most profound influence was the introduction of quantum logic. The possibility of a quantum Turing machine was initially suggested by David Deutsch (1985). At the same time Richard Feynman (1982), faced with the difficult problem of simulating quantum systems with classical computers, suggested the possibility of building a powerful computer out of quantum systems that would easily outperform classical computers. The idea of quantum computation was born, which opened up numerous potential applications.

Conceptually, quantum computers follow the paradigm of their classical counterparts, but with some distinct modifications. In place of the bit that describes the classical state 0 or 1 of a binary system, there is the qubit. The qubit encodes the state of a binary quantum system, which is in general a linear superposition of the states $|0\rangle$ and $|1\rangle$. Quantum mechanics also allows superpositions between states of many qubits, the so-called entangled states. The presence of entanglement dramatically increases the dimension of the encoding space. Qubits are evolved by quantum gates. These are unitary operations that can perform any desired state transformation. A consistent computational model can be made out of qubits and quantum gates that add intriguing new possibilities to classical computation.

The main obstacle for practical applications of quantum computers is to physically realise them in the laboratory. For a computation to be reliable one needs to be very accurate in initialising qubits, performing quantum gates and reading the computational outcome. Errors may enter the system in the form of control inaccuracies inherent to any experimental procedure or as uncontrollable perturbations from the environment of the computer. Quantum error correction has been developed (Shor, 1995; Steane, 1996) to overcome this problem. It allows one, at least theoretically, to correct errors and perform meaningful quantum information processing. Still, the current thresholds for reliable quantum computation are much higher than what is currently possible, even with state-of-the-art experiments. Hence, there is an active search for novel quantum computation architectures that resolve the problem of errors in a more efficient way than current models.

Motivated by the richness of quantum physics, a huge number of quantum computing schemes have already been proposed. Novel computational models appeared which are based on measurements (Raussendorf and Briegel, 2001), on adiabatic transitions (Farhi *et al.*, 2001), on geometric phases (Zanardi and Rasetti, 1999) or on topological evolutions (Kitaev, 2003). Each scheme has its own merits and drawbacks. Diverse as they might look, all models are computationally equivalent to the quantum Turing machine. The topological quantum computation scheme will be presented in the next chapter. In this chapter we lay down the principles of quantum computation and we review some commonly used computational schemes.

3.1 Qubits and their manipulations

3.1.1 Quantum bits

Let us first focus on the encoding elements of quantum information, the qubits. The crucial characteristic of qubits is that their states $|0\rangle$ and $|1\rangle$ can be prepared in any linear superposition of the form

$$|\psi\rangle = a_0 |0\rangle + a_1 |1\rangle, \tag{3.1}$$

where a_0 and a_1 are complex numbers with $|a_0|^2 + |a_1|^2 = 1$. Equivalently, a 'qudit' employs d states $\{|0\rangle, \ldots, |d-1\rangle\}$ to encode one element of information. When a qubit is measured, then the state $|\psi\rangle$ collapses to its components along the measurement basis. If the measurement is on the $\{|0\rangle, |1\rangle\}$ basis then the outcome is $|0\rangle$ with probability $|a_0|^2$ and $|1\rangle$ with probability $|a_1|^2$. The possibility of employing superpositions is a first point of departure from classical computers. Such superpositions are responsible for the efficiency of certain quantum algorithms compared to their classical counterparts. Another difference is the way qubits are composed together. The general state of n qubits is described by

$$|\psi\rangle = \sum_{i_1,i_2,\ldots,i_n=\{0,1\}^n} a_{i_1,i_2,\ldots,i_n} |i_1, i_2, \ldots, i_n\rangle, \tag{3.2}$$

where the $a_{i_1,i_2,\ldots,i_n}$'s are complex coefficients with $\sum_{i_1,i_2,\ldots,i_n=\{0,1\}^n} |a_{i_1,i_2,\ldots,i_n}|^2 = 1$. Here the tensor product between states is employed such that $|i_1,i_2,\ldots,i_n\rangle = |i_1\rangle \otimes |i_2\rangle \otimes \ldots \otimes |i_n\rangle$). The summation in (3.2) runs over 2^n numbers, thereby giving rise to an exponentially large Hilbert space, while the classical encoding space is only $2n$-dimensional. Such an increase in the number of computational states can provide new shortcuts in the algorithmic evolutions from the input of computation to the desired answer, thus speeding up the computation.

Quantum mechanically it is possible to have states of matter that can be described by entangled states. These are superposition states that describe composite systems, such as the ones given in (3.2). An example, is the maximally entangled state between two qubits A and B

$$|\psi_{AB}\rangle = \frac{1}{\sqrt{2}}\big(|0_A 0_B\rangle + |1_A 1_B\rangle\big), \tag{3.3}$$

known as a Bell state. Measuring qubit A reveals instantaneously the state of qubit B even if the two qubits are positioned an arbitrarily large distance apart. This correlation goes beyond the probabilistic scenario met in classical physics and is spectacularly witnessed by the violation of the Bell inequalities (Bell, 1966). Entangled states give rise to many dramatic effects, ranging from the double-slit experiment (Feynman $et\ al.$, 1963) to the Einstein–Podolsky–Rosen paradox (Einstein $et\ al.$, 1935).

3.1.2 Decoherence and mixed states

When the quantum state of a system is not exactly known, then classical probabilities mix with the amplitudes of the quantum states. This can be the result of quantum decoherence, due to the interaction of a quantum system with its environment. The system combined with the environment can be described non-probabilistically and is subject to unitary evolutions. However, ignoring the environment leads to effective non-unitary evolutions of the system alone. In this case the system cannot be described by pure states such as the one given in (3.2), but by density matrices.

Consider an orthonormal basis set of states $\{|\psi_i\rangle\}$ of a system. Then its density matrix can be written as

$$\rho = \sum_i p_i |\psi_i\rangle \langle \psi_i|, \tag{3.4}$$

which implies that the state $|\psi_i\rangle$ occurs with probability p_i. The normalisation of the probabilities, $\sum_i p_i = 1$, implies $\mathrm{tr}(\rho) = 1$. The expectation value of an operator O with respect to a density matrix ρ is given by

$$\langle O \rangle_\rho = \mathrm{tr}(\rho O) = \sum_i p_i \langle \psi_i | O | \psi_i \rangle. \tag{3.5}$$

Density matrices provide the most general description of a quantum system. For example, a system in a pure state $|\psi\rangle$ has the density matrix

$$\rho = |\psi\rangle \langle \psi|. \tag{3.6}$$

If the density matrix cannot be written as in (3.6), then the system is in a mixed state. The maximally mixed state of a qubit is given by

$$\rho = \frac{1}{2}\,|0\rangle\,\langle 0| + \frac{1}{2}\,|1\rangle\,\langle 1|. \tag{3.7}$$

A mixed state reveals our ignorance about the quantum state of a system and can arise in various ways. To illustrate this we now consider two qubits in the entangled state (3.3), where A plays the role of a system and B the role of an environment. Assume you cannot access qubit B at all. Mathematically, this corresponds to wiping out the information from B by performing the following steps. First we create the density matrix of the composite system $\rho_{AB} = |\psi_{AB}\rangle\,\langle\psi_{AB}|$ and then we trace out the Hilbert space of B. The latter corresponds to partially tracing the density matrix with respect to the indices of the B system, giving finally

$$\rho_A = \mathrm{tr}_B(\rho_{AB}) = \frac{1}{2}\,|0_A\rangle\,\langle 0_A| + \frac{1}{2}\,|1_A\rangle\,\langle 1_A|. \tag{3.8}$$

The tracing procedure resulted in the maximally mixed state (3.7) for qubit A. Ignoring part of a system leads in general to a mixed state.

Commonly, the environment is described by a very large Hilbert space, so its state is considered to be inaccessible. Hence, a system that interacts with the environment is in general described by a mixed state. If a system with Hamiltonian H is in thermal equilibrium with its environment with temperature T, then the density matrix of the system is given by

$$\rho = \frac{e^{-\frac{H}{kT}}}{\mathrm{tr}(e^{-\frac{H}{kT}})}, \tag{3.9}$$

known as the thermal state, where k is the Boltzmann constant. This is a mixed state that becomes pure only in the limit $T \to 0$ (see Exercise 3.1).

If the computational state becomes mixed either due to interactions with the environment or due to lack of detailed knowledge about our control procedure, then the quantum computation can give erroneous results. The aim of topological quantum computation is to keep the computational states as close as possible to the desired pure quantum states.

3.1.3 Quantum gates and projectors

The processing of the encoded quantum information is usually performed by quantum gates. These are reversible quantum evolutions that operate on one, two or more qubits simultaneously, in the same way as classical gates do. As quantum evolutions are described by unitary matrices, quantum gates between n qubits are elements of the unitary group $U(2^n)$. For example, one-qubit gates include the Pauli matrices

$$\sigma^x = \begin{pmatrix} 0 & 1 \\ 1 & 0 \end{pmatrix}, \ \sigma^y = \begin{pmatrix} 0 & -i \\ i & 0 \end{pmatrix}, \ \sigma^z = \begin{pmatrix} 1 & 0 \\ 0 & -1 \end{pmatrix}. \tag{3.10}$$

Here σ^x is known as the classical NOT gate that changes the input 0 or 1 to the output 1 or 0, respectively. A more intriguing gate is the Hadamard gate given by

$$H = \frac{1}{\sqrt{2}} \begin{pmatrix} 1 & 1 \\ 1 & -1 \end{pmatrix} \qquad (3.11)$$

that transforms $|0\rangle$ into the quantum superposition $(|0\rangle + |1\rangle)/\sqrt{2}$ and $|1\rangle$ into $(|0\rangle - |1\rangle)/\sqrt{2}$. This operation corresponds to a rotation of the original basis by $\pi/4$ around the z axis.

To create entanglement between two qubits we need to introduce two-qubit quantum gates. An important class of two-qubit gates is the controlled gates, CU. These gates treat one qubit as the controller and the other one as the target. The action of CU is to leave the target qubit unaffected when the control is in state $|0\rangle$ and to apply the unitary matrix U on the target qubit when the control is in state $|1\rangle$. An example of such a gate is the controlled-NOT gate, also known as CNOT, where $U = \sigma^x$. It is given explicitly by

$$\text{CNOT} = \begin{pmatrix} 1 & 0 & 0 & 0 \\ 0 & 1 & 0 & 0 \\ 0 & 0 & 0 & 1 \\ 0 & 0 & 1 & 0 \end{pmatrix}. \qquad (3.12)$$

The controlled-phase gate, also known as CP, has $U = \sigma^z$ and is given by

$$\text{CP} = \begin{pmatrix} 1 & 0 & 0 & 0 \\ 0 & 1 & 0 & 0 \\ 0 & 0 & 1 & 0 \\ 0 & 0 & 0 & -1 \end{pmatrix}. \qquad (3.13)$$

One can show that all CU gates are unitary when U is a unitary matrix. The controlled character of these gates makes them capable of generating entangled states. For example, applying a CNOT to the unentangled two-qubit state $(|0\rangle + |1\rangle)/\sqrt{2} \otimes |0\rangle$ gives

$$\text{CNOT} \left(\frac{|0\rangle + |1\rangle}{\sqrt{2}} \otimes |0\rangle \right) = \frac{1}{\sqrt{2}} (|00\rangle + |11\rangle), \qquad (3.14)$$

which is a maximally entangled state of two qubits. Another often considered gate is the SWAP gate, which exchanges the states of two qubits, such that $\text{SWAP}|ij\rangle = |ji\rangle$. The SWAP gate is not an entangling gate since it can also be performed in classical bits but is still very useful in quantum algorithms. Multi-qubit gates are also possible with many controlled qubits and one or more target qubits. Usually, when we say that we can realise a specific set of gates, we assume that we are able to apply each of them to any qubits we want.

Beyond unitary evolutions, the manipulation of quantum information can include projectors. These are a set of operators $\{P_i\}$ that square to themselves and the product of two different projectors P_i and P_j can be chosen to be zero:

$$P_i^2 = P_i \text{ and } P_i P_j = 0 \text{ for } i \neq j. \qquad (3.15)$$

When a projector acts on a certain quantum state it gives back its component within a certain subspace of the Hilbert space. An application of projectors is to employ them as a mathematical tool for the measurement of qubits. For example, the operator

$$P_0 = |0\rangle \langle 0| = \begin{pmatrix} 1 & 0 \\ 0 & 0 \end{pmatrix} \tag{3.16}$$

projects the general qubit state $|\psi\rangle = a|0\rangle + b|1\rangle$ to $|0\rangle$ with probability $|a|^2 =$ tr$(|\psi\rangle \langle\psi| P_0)$. The projector onto the state $|1\rangle$ is given by $P_1 = \mathbb{1}_2 - P_0$, where $\mathbb{1}_2$ is the two-dimensional identity matrix. A measurement along a general direction of the one-qubit state space can be obtained by the projector $P = |\psi\rangle \langle\psi|$ with $|\psi\rangle = \cos\theta |0\rangle + e^{i\phi} \sin\theta |1\rangle$. The identity operator can be decomposed into a sum of projection operators

$$\mathbb{1} = \sum_{n=1}^{N} |\psi_n\rangle \langle\psi_n|, \tag{3.17}$$

where $\{|\psi_n\rangle, n = 1, \ldots, N\}$ is a complete orthonormal set of basis states.

The projection space of states does not need to be one-dimensional. For example, the operator $P = |0\rangle \langle 0| + |1\rangle \langle 1|$ projects the quantum state $|\psi\rangle = a|0\rangle + b|1\rangle + c|2\rangle$ with $|a|^2 + |b|^2 + |c|^2 = 1$ onto the two-dimensional subspace described by the general state $|\psi'\rangle = (a|0\rangle + b|1\rangle)/\sqrt{|a|^2 + |b|^2}$. Projections of a many-qubit state onto an entangled two-qubit state can also be considered. For example, $P = |\psi\rangle \langle\psi|$ with $|\psi\rangle = (|00\rangle + |11\rangle)/\sqrt{2}$ projects any two-qubit state onto this maximally entangled state.

Another application of projectors is to define Hamiltonians with very specific properties. Assume we want to build a Hamiltonian that has the state $|\psi_0\rangle$ as its ground state. Suppose we know a set of Hermitian projectors P_i, $i = 1, \ldots, k$ that project to different subspaces that are not orthogonal to each other. If all of these subspaces share a single common state $|\psi_0\rangle$, i.e., $P_i |\psi_0\rangle = |\psi_0\rangle$ for all i, then the Hamiltonian

$$H = -\sum_{i=1}^{k} (\mathbb{1} - P_i) \tag{3.18}$$

has $|\psi_0\rangle$ as its ground state with eigenvalue $E_0 = 0$. Such a description will be employed in Chapter 5 to present a certain topological model.

3.2 Quantum circuit model

Let us now present the basic characteristics of the quantum circuit model. This computational model employs a sequence of quantum gates acting on a series of qubits to transform their initial state into the desired output. The total unitary evolution that acts on the qubits is called the algorithm. If we have at our disposal a so-called complete set of quantum gates, then we can realise an arbitrary quantum algorithm. An example of a circuit model that consists of one- and two-qubit gates is given in Figure 3.1. There, all qubits are initialised in the state $|0\rangle$. The boxes symbolise gates while the lines indicate the qubits on which a gate acts. A measurement of all the qubits at the end of the computation reveals the outcome. Expressing an algorithm in terms of basic quantum gates makes it easy to determine its complexity.

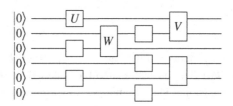

Fig. 3.1 An example of a circuit model, where time evolves from left to right. Qubits are initially prepared in state $|0\rangle$ and one- and two-qubit gates, such as U, V and W, act on them in succession. Boxes attached on a single line correspond to single-qubit gates while boxes attached on two lines correspond to two-qubit gates.

3.2.1 Quantum algorithm and universality

A specific quantum algorithm U applied on n qubits is an element of the unitary group $U(2^n)$. It acts on an initially prepared quantum state $|\psi_0\rangle$ that encodes the input of the problem. Its output state $|\psi\rangle = U|\psi_0\rangle$ encodes the solution of the problem. The algorithm needs to be designed such that the information encoded in $|\psi\rangle$ can be read by projective measurements. Usually, we take $|00\ldots0\rangle$ as the initial state and the encoding of input step $|\psi_0\rangle = U_0|00\ldots0\rangle$ is considered part of the algorithm. The unitary matrix U_0 depends on the information we want to encode, while the algorithm U is independent of the input of the computation. It depends only on the number, n, of employed qubits.

To realise a given algorithm we would like to break it down into smaller elements that are physically easier to implement. Commonly, these smaller elements are taken to be one- and two-qubit gates that can be applied to any desired qubit at any time. Composed together in a temporal fashion they give rise to the circuit model of quantum computation.

A natural question arises: which types of quantum gates are needed to be able to perform any given algorithm? For example, one could try to employ as few different types of quantum gates as possible, acting only on a small number of qubits at the time, e.g., one or two. A finite set of quantum gates that can efficiently generate any given unitary matrix is called universal. Universality is an important property that needs to be satisfied from any implementation scheme of a quantum computer. Several universal sets of quantum gates are known. The simplest one comprises arbitrary one-qubit gates and a maximally entangling two-qubit gate like the CNOT gate.

An alternative way to formulate the universality condition, without resorting to the notion of qubits, is the following. We want to find a small set of unitary matrices that can reproduce an arbitrary given element of $U(N)$. This can be achieved by taking two unitary matrices of $U(N)$ and calculating their product. By multiplying them, a third, independent unitary matrix is produced. This process can be iterated several times, where each time we allow any unitary matrix to be employed from the previous steps. If the two initial matrices are correctly chosen, then the output of the procedure covers densely the whole unitary group $U(N)$. An arbitrary unitary matrix in $U(N)$, e.g., corresponding to an algorithm, can thus be produced with a controllable error (Kitaev, 1997). If a physical system supports a set of unitary matrices that can span $U(N)$ densely, then it can operate as a universal quantum computer. This description is of particular interest to topological quantum computation, where statistical evolutions provide the available set of unitary matrices. It can thus determine if a given anyonic model is universal or not (Freedman *et al.*, 2002a, b).

To date, several quantum algorithms are known that have the potential to outperform the corresponding classical algorithms. Among them, two algorithms are most prominently shaping the landscape of quantum computation. These are the factoring algorithm by Peter Shor (1997) and the searching algorithm by Lov Grover (1996). The factoring algorithm determines the prime factors of a given integer. This is an important algorithm as many modern secure data encryption protocols are based on the fact that there is no known efficient classical algorithm for this task. By efficient we mean that the amount of resources, e.g., the number of gates that need to be performed during the computation, increases only polynomially with the size of the input. The amount of resources of all known classical factoring algorithms increases exponentially with the size of the number we want to factor. Hence we are currently unable to factor arbitrarily large numbers. Shor (1997) invented a quantum algorithm that would perform this task efficiently, thus making the construction of a quantum computer highly desirable.

Classically, searching an unsorted database by going through all its elements takes a time that increases linearly with the size of the database. Grover (1996) proved that a quadratic speedup is obtained if the database is searched by a quantum computer. While this does not constitute an exponential speedup as in the case of the factoring algorithm, it is known that we cannot ever build a classical algorithm that can do better than 'linear'. Moreover, it is known that the quadratic speedup of Grover's algorithm is actually optimal.

There are more quantum algorithms that perform several tasks, but none of them appears yet to have as general an impact as the factoring or searching algorithms. In Chapter 8 we present a quantum algorithm (Freedman *et al.*, 2003) that outperforms its classical counterpart in the task of evaluating Jones polynomials. The latter are topological invariants of links and knots that have several applications, albeit specialised.

3.2.2 Computational complexity

From the previous analysis it is easy to imagine that employing quantum logic might actually give rise to faster algorithms compared with the classical ones. But how much faster can they be? Among other things, information theory deals with the classification of various problems in different complexity classes (Bennett, 1982; Nielsen and Chuang, 2000). These are determined by the speed we can solve them. Essentially, there are two classes of problems: the ones that can be solved in polynomial time as a function of the input size of the problem, and the ones that need an exponential amount of time. The latter problems become computationally intractable even for moderate input sizes. Problems that can be solved in polynomial time are called easy, while the ones that require exponential time are called hard.

The simplest complexity class, denoted by **P**, consists of the problems that can be solved polynomially fast with classical computers. Similarly, **NP** denotes the class of problems for which one can efficiently verify if a given statement is a solution to the problem or not. All

Table 3.1 Computational complexity of algorithmic problems	
P	Polynomially easy to solve
NP	Polynomially easy to verify solution
BQP	Polynomially easy to solve with quantum computer

P problems are also **NP**, as finding a solution means that you can also verify it efficiently. The converse is not always true. In particular, all known algorithms strongly suggest that there are elements in **NP** that do not belong in **P**. In other words, there exist problems that are computationally hard to solve on a classical computer (i.e., they need an exponential amount of resources), even though a known solution could be verified in polynomial time.

The invention of quantum computers introduced a new class of problems, called **BQP**. This class includes all problems that can be solved easily on a quantum computer. The characterisation of this class of problems with respect to their classical counterparts has proved challenging. For example, it is not yet known if there are non-**P** problems (i.e., classically exponentially hard problems) that belong in **BQP** (i.e., they can easily be solved on a quantum computer). A proof of their existence would establish quantum computing as a new computational paradigm in terms of efficiency. Indeed, we do not know for sure whether or not there exists a classical **P** algorithm that can factor efficiently. Also, the existence of a quantum searching algorithm does not change the complexity class of the corresponding classical algorithm, since it does not provide an exponential speedup. Table 3.1 summarises the main computational complexity classes.

It is surprising that, even though a significant amount of research is dedicated to quantum algorithms, very few (albeit important) algorithms exist that have a wide applicability. Moreover, none exists that could shift the complexity class of **BQP** to include **NP**-hard problems. This might be due to our lack of hands-on experience when it comes to dealing with the quantum world. Nevertheless, it is known that quantum computers can perform certain useful tasks exponentially faster than classical ones, like factoring large numbers or simulating complex quantum systems. Hence, the need for constructing full-scale quantum computers is much more than just a curiosity. It could change the way we process information, with the potential for a wide range of applications in security, technology, science and entertainment.

3.3 Other computational models

The quantum circuit model offers a straightforward architecture for building a quantum computer. It can be implemented by turning on and off designed interactions with time evolutions that correspond to quantum gates. Interestingly, this is not the only possible way to process quantum information. The desired algorithm, given in terms of a unitary matrix U, can also be realised in a number of different ways. This plurality of implementations stems from the variety of exotic effects harboured by quantum mechanics. In the following we describe three alternative ways to process information, namely one-way quantum computation (Raussendorf and Briegel, 2001), adiabatic quantum computation (Farhi *et al.*, 2001) and holonomic quantum computation (Zanardi and Rasetti, 1999). Although these methods may seem fundamentally different from the circuit model, they are all computationally equivalent to it. Our interest in these models is due to their similarities with topological quantum computation or due to the possibility of combining them with topological manipulations, thereby resulting in new computational paradigms.

3.3.1 One-way quantum computation

The starting point and initial resource of one-way or measurement-based quantum computation (Raussendorf and Briegel, 2001) is a highly entangled state between qubits, the cluster state. Once the resource has been prepared, performing an algorithm requires only single-qubit measurements and classical processing of the measurement outcomes. The underlying principle of this quantum computing method is the following: it realises a reversible evolution of an encoded Hilbert space by measurements on the qubits that are prepared in the cluster state. When these non-reservable measurements are performed in a clever way, they induce a unitary evolution of the encoded qubits.

Let us start with the description of the cluster state $|\psi_c\rangle$. Consider qubits residing at the vertices of a square lattice. Take the complete set of Hermitian operators

$$T_i = \sigma_i^x \sigma_{i_1}^z \sigma_{i_2}^z \sigma_{i_3}^z \sigma_{i_4}^z, \tag{3.19}$$

where i is a site of the lattice and $i_1, \ldots, i_4$ are its four neighbouring vertices, as illustrated in Figure 3.2(a). The operators T_i are modified at the boundaries so that the operators that correspond to vertices outside the lattice are absent. Since all T_i's commute with each other, it is possible to find a common set of eigenstates by diagonalising each of them separately. As $T_i^2 = 1$ their eigenvalues are ± 1. The cluster state $|\psi_c\rangle$ is hence uniquely defined by the condition

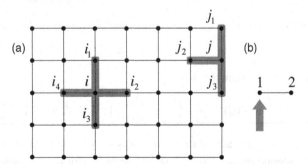

Fig. 3.2 (a) A cluster state can be defined on a square lattice with qubits at its vertices. The stabiliser operator $T_i = \sigma_i^x \sigma_{i_1}^z \sigma_{i_2}^z \sigma_{i_3}^z \sigma_{i_4}^z$ is defined to act on i and on its neighbouring qubits. A reduced version of the stabiliser operator is defined on the boundary of the system, e.g., $T_j = \sigma_j^x \sigma_{j_1}^z \sigma_{j_2}^z \sigma_{j_3}^z$. (b) Qubits 1 and 2 are depicted that are prepared in a two-qubit cluster state. When we perform a measurement on qubit 1 along a particular axis then qubit 2 is evolved in a non-trivial way.

$$T_i \,|\,\psi_c\rangle = |\,\psi_c\rangle \,, \text{ for all } i. \tag{3.20}$$

A useful property of $|\,\psi_c\rangle$ is that all of its qubits are maximally entangled with their neighbours. If we measure one of the qubits of the cluster state, then this qubit is removed from the state while the rest remain entangled with each other. Actually, each measurement on boundary qubits along an arbitrary axis transforms all other qubit states in a non-trivial way. When the measurement outcome is known, then we can write this transformation as a unitary evolution. Consider a rectangular cluster state, such as the one shown in Figure 3.2(a). Suppose we perform measurements on each column of qubits from left to right, thus simulating the passage of time. It has been shown that particular patterns of measurements can give rise to single-qubit rotations and the controlled-NOT gate (Raussendorf and Briegel, 2001). This makes one-way quantum computation equivalent to the circuit model.

A simple example that demonstrates the principle of one-way quantum computation is the smallest possible non-trivial cluster state that contains only two qubits, shown in Figure 3.2(b). Its stabilisers are

$$T_1 = \sigma_1^x \sigma_2^z \text{ and } T_2 = \sigma_1^z \sigma_2^x. \tag{3.21}$$

The state which satisfies $T_1 \,|\,\psi_c\rangle = |\,\psi_c\rangle$ and $T_2 \,|\,\psi_c\rangle = |\,\psi_c\rangle$ is given by

$$|\,\psi_c\rangle = \frac{1}{\sqrt{2}}\big(|+0\rangle + |-1\rangle\big), \tag{3.22}$$

where $|\pm\rangle = (|0\rangle \pm |1\rangle)/\sqrt{2}$. Let us measure qubit 1 in a rotated basis $|0(\theta,\phi)\rangle = U(\theta,\phi)\,|0\rangle$ and $|1(\theta,\phi)\rangle = U(\theta,\phi)\,|1\rangle$, where

$$U(\theta,\phi) = \begin{pmatrix} \cos\theta & e^{i\phi}\sin\theta \\ e^{-i\phi}\sin\theta & -\cos\theta \end{pmatrix} \tag{3.23}$$

is an arbitrary SU(2) matrix. The angles θ and ϕ determine the direction of the measurement. One can show that the measurement outcome $|0\rangle$ on the first qubit results in the state

$$|\,\psi_0\rangle = P(\phi)R(\theta)\,|0\rangle \tag{3.24}$$

for the second qubit, while the outcome $|1\rangle$ results in the state

$$|\,\psi_1\rangle = \sigma^z P(-\phi)\sigma^x R(\theta)\,|1\rangle \tag{3.25}$$

for the second qubit, where

$$P(\phi) = \frac{1}{\sqrt{2}}\begin{pmatrix} 1 & e^{i\phi} \\ 1 & -e^{i\phi} \end{pmatrix} \text{ and } R(\theta) = \begin{pmatrix} \cos\theta & \sin\theta \\ \sin\theta & -\cos\theta \end{pmatrix}. \tag{3.26}$$

The above example shows that it is possible to implement any single-qubit rotation on the second qubit of the cluster state by performing a measurement in an appropriately chosen basis on the first qubit. This reverse engineering generates $P(\phi)$ and $R(\theta)$ unitary rotations, up to spurious Pauli rotations that can be easily dealt with at the end of the computation. A similar principle applies to measurements of a large two-dimensional cluster state that can produce arbitrary one- and two-qubit gates. Hence, universality of one-way quantum computation can be demonstrated by drawing its analogy to the circuit model. Although being essentially equivalent, cluster state quantum computation is desired over the circuit model when the preparation of a cluster state and single-qubit measurements are physically easier to perform than the successive application of entangling quantum gates.

3.3.2 Adiabatic quantum computation

In adiabatic quantum computation (Farhi *et al.*, 2001) the algorithm is produced by an adiabatic process. To see how this is possible, consider a Hamiltonian with a non-degenerate ground state and a well-defined energy gap above it. We assume now that the Hamiltonian acts on a set of particles that encode qubits and that we can change the Hamiltonian in a controlled way. In this scheme information is encoded in the ground state of the Hamiltonian. For this encoding to be meaningful we assume that there is a finite energy gap above the ground state at all times. Processing of information is performed by slowly changing the Hamiltonian parameters so that, due to adiabaticity, the system remains in its unique ground state. The condition that ensures adiabaticity is that the kinetic energy that corresponds to the speed with which we change the parameters of the Hamiltonian is much smaller than the energy gap above the ground state. So, transitions to excited states are suppressed and the system is well described at all times by its ground state. The manipulation of the Hamiltonian is performed so that the initial ground state can be prepared easily while the final ground state is the desired solution of the problem in question.

To be more concrete, consider the Hamiltonian

$$H(\lambda) = (1 - \lambda)H_i + \lambda H_f, \tag{3.27}$$

where the parameter $\lambda \epsilon [0, 1]$ is monotonic in time, e.g., $\lambda = t/T$. The initial Hamiltonian H_i is a simple one with a known ground state $|\psi_i\rangle$. The final Hamiltonian H_f is designed so that its unique ground state, $|\psi_f\rangle$, is the solution of the computational problem at hand. Even when the state $|\psi_f\rangle$ is hard to evaluate, the Hamiltonian H_f could easily be constructible. For example, if we are interested in a satisfiability problem we can energetically penalise every configuration that violates any of the conditions we want to satisfy. Hence, the ground state of H_f will violate the least number of such conditions.

Adiabatic quantum computation gives a way to obtain $|\psi_f\rangle$ from an adiabatic evolution. To achieve this we require that $H(\lambda)$ remains gapped at all times, i.e., the energy splitting between the ground state and the excited state does not become zero for any λ. At $t = 0$ we prepare the state of the system to be its ground state, i.e., $|\psi(0)\rangle = |\psi_i\rangle$. Subsequently, we change the parameters of the Hamiltonian for time T, obtaining finally the state $|\psi(T)\rangle$. We take T to be large enough compared to the minimal gap of the Hamiltonian so that the adiabatic approximation holds and the final state $|\psi(T)\rangle$ is very close to the target state $|\psi_f\rangle$.

As an example, consider a two-qubit system subject to the Hamiltonian (3.27) with

$$H_i = -\sigma_1^z - \sigma_2^z \tag{3.28}$$

and

$$H_f = -\sigma_1^z \sigma_2^z - \sigma_1^x \sigma_2^x. \tag{3.29}$$

The ground state of the initial Hamiltonian, H_i, is

$$|\psi_i\rangle = |00\rangle \tag{3.30}$$

and the ground state of the final Hamiltonian, H_f, is

$$|\psi_f\rangle = (|00\rangle + |11\rangle)/\sqrt{2}. \tag{3.31}$$

By slowly increasing λ from 0 to 1 compared with the smallest energy gap of $H(\lambda)$, a maximally entangled state $|\psi_f\rangle$ is obtained out of a product state $|\psi_i\rangle$. During this process the Hamiltonian remains always gapped, i.e., there is a non-zero energy gap ΔE between the ground state and the first excited state, as shown in Figure 3.3(a).

In adiabatic quantum computation the computational complexity of a problem is mainly given in terms of the overall time of the evolution, T. In particular, the evolution time $T(n)$ depends on the number, n, of qubits the Hamiltonian $H(\lambda)$ acts on. It is expected that the larger the number n the longer the evolution will last. By studying specific examples like Grover's searching algorithm (Roland and Cerf, 2002) it is possible to see that at certain points of the evolution the energy gap becomes small, as shown in Figure 3.3(a). To satisfy the adiabaticity condition the evolution needs to be slower in this neighbourhood, thus increasing the overall time $T(n)$. When n tends to infinity then the gap goes to zero for some value of λ. This λ constitutes a quantum critical point, as shown in Figure 3.3(b). By employing more than one control parameter, it is possible to avoid critical regions and to construct an evolution that does not need to slow down to remain adiabatic. Nevertheless, the path becomes longer so a longer time needs to be spent to traverse it.

It is an amazing fact that adiabatic quantum computation is actually equivalent to the circuit model (Aharonov, 2007). Known quantum algorithms can be translated to Hamiltonian (3.27), where the computation is realised in terms of an adiabatic evolution rather than in terms of precisely timed interactions. As we shall see later, topological quantum computation resembles an adiabatic quantum computation with constant energy gap, where the quasiparticle coordinates provide the control parameters of the Hamiltonian.

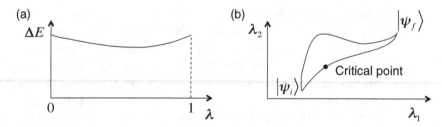

Fig. 3.3 (a) The energy gap ΔE of a simple two-qubit Hamiltonian $H(\lambda) = (1 - \lambda)(-\sigma_1^z - \sigma_2^z) + \lambda(-\sigma_1^z \sigma_2^z - \sigma_1^x \sigma_2^x)$ that takes the product state $|00\rangle$ to the entangled state $(|00\rangle + |11\rangle)/\sqrt{2}$ when λ changes adiabatically from 0 to 1. More complex problems give rise to smaller minimum energy gaps, thus forcing the adiabatic evolution to slow down. (b) Different paths can be taken in the parametric space $\{\lambda_1, \lambda_2\}$ with initial and final points that correspond to $|\psi_i\rangle$ and $|\psi_f\rangle$, respectively. These paths may be short, passing near critical regions, or long, avoiding any criticality.

3.3.3 Holonomic quantum computation

We have seen in Chapter 2 that Abelian (Berry, 1984) and non-Abelian (Wilczek and Zee, 1984) geometric phases describe certain evolutions of quantum systems in terms of geometric means. In this subsection we show how one can employ these evolutions to perform quantum computation (Zanardi and Rasetti, 1999). We start with an energy degenerate quantum system subject to adiabatic cyclic evolutions in some parametric space of a Hamiltonian. The corresponding logical state space consists of degenerate states. The quantum logical gates are given in terms of non-Abelian geometric phases acting on this space. In order for such a system to support qubits, the degenerate subspace needs to have a natural tensor product structure. Moreover, the control space needs to be rich enough such that arbitrary geometric evolutions can be built, resulting in universality.

One-qubit holonomic gates can be realised with a three-level system, $|\alpha\rangle$, $\alpha = 0, 1, 2$ subject to the Hamiltonian

$$H(\mathbf{z}) = \mathcal{U}(\mathbf{z})H_0\mathcal{U}(\mathbf{z})^\dagger \quad \text{with } H_0 = \begin{pmatrix} 0 & 0 & 0 \\ 0 & 0 & 0 \\ 0 & 0 & 1 \end{pmatrix}. \tag{3.32}$$

Non-trivial isospectral transformations of H_0 are parameterised by the U(2) rotations between states $|0\rangle$ and $|2\rangle$ as well as between $|1\rangle$ and $|2\rangle$, as shown in Figure 3.4(a). A general transformation is given by the unitary rotation $\mathcal{U}(\mathbf{z}) = U_1(z_1)U_2(z_2)$, with $U_\alpha(z_\alpha) = \exp(z_\alpha|\alpha\rangle\langle 2| - \bar{z}_\alpha|2\rangle\langle\alpha|)$ for $\alpha = 0, 1$. The complex parameter z_α may be decomposed as $z_\alpha = \theta_\alpha \exp i\phi_\alpha$. As we have seen in Subsection 2.2.2, when loops C are adiabatically spanned in the parametric space of this system, we obtain the holonomies

$$\Gamma_\mathbf{A}(C) = \mathbf{P}\exp\oint_C \mathbf{A} \cdot d\lambda, \quad \text{where} \quad (A^\mu)_{\alpha\beta} = \langle\alpha|\mathcal{U}^\dagger(\lambda)\frac{\partial}{\partial\lambda^\mu}\mathcal{U}(\lambda)|\beta\rangle. \tag{3.33}$$

Here $\mathbf{P}$ is the path-ordering symbol, $\alpha, \beta = 0, 1$ parameterise degenerate states and $\lambda^\mu \in \{\theta_1, \theta_2, \phi_1, \phi_2\}$. The connection $\mathbf{A}$ is a vector whose components are matrices. As it is irreducible, the holonomy $\Gamma_\mathbf{A}(C)$ can generate the whole group U(2).

We now want to realise specific gates out of holonomic evolutions. If, for example, we want to implement $U \in$ U(2), we have to find the loop C such that $\Gamma_\mathbf{A}(C) = U$. To find the appropriate C we perform the following analysis. The loop integral

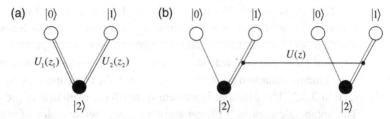

Fig. 3.4 (a) A diagram that represents the transformations of $H(\mathbf{z})$. Three states are depicted together with the two unitary rotations $U_1(z_1)$ and $U_2(z_2)$ that give rise to holonomic one-qubit gates. (b) The composite rotation $U(z)$ between two subsystems gives rise to a two-qubit gate.

$$\oint_C \mathbf{A} \cdot d\boldsymbol{\lambda} = \oint_C A_{\lambda^1} d\lambda^1 + A_{\lambda^2} d\lambda^2 + A_{\lambda^3} d\lambda^3 + \cdots \qquad (3.34)$$

is the main ingredient of the holonomy. Due to the path-ordering symbol $\mathbf{P}$ it is not possible to just calculate it and evaluate its exponential as the connection components do not commute with each other in general. Still, it is possible to consider particular loop configurations that bypass this problem. First, choose C such that it lies on one plane (λ^1, λ^2). So only two components of $\mathbf{A}$ are involved. Second, choose the position of the plane so that $\mathbf{A} \cdot d\boldsymbol{\lambda} = A^{\lambda^2} d\lambda^2$, i.e., the connection restricted on this plane has $A^{\lambda^1} = 0$. Then the two components of $\mathbf{A}$ commute with each other and the path-ordering symbol can be dropped. Still the related curvature, $F_{\mu\nu} = \partial_\mu A_\nu - \partial_\nu A_\mu + [A_\mu, A_\nu]$, can be non-vanishing, thereby giving rise to a non-trivial holonomy.

To apply this approach to Hamiltonian (3.32) we choose the following loops with the corresponding holonomies. For $C_1 \in (\theta_a, \phi_a)$ with $a = 1, 2$ we have

$$\Gamma_{\mathbf{A}}(C_1) = \exp(-i\Sigma_1 \sigma_\alpha^3), \qquad (3.35)$$

where $\sigma_\alpha^3 = |\alpha\rangle\langle\alpha|$ with $\alpha = 0, 1$ and where Σ_1 is the area of the surface the path C_1 encloses when projected on a sphere with coordinates $2\theta_a$ and ϕ_a. For $C_2 \in (\theta_1, \theta_2)$ with $\phi_1 = 0$ and $\phi_2 = 0$ we have

$$\Gamma_{\mathbf{A}}(C_2) = \exp(-i\Sigma_2 \sigma^2) \qquad (3.36)$$

with $\sigma^2 = -i|0\rangle\langle 1| + i|1\rangle\langle 0|$. Here Σ_2 is the area on the sphere with coordinates θ_1 and θ_2. Hence, we have enough freedom to produce any arbitrary one-qubit gate.

A set of m such subsystems gives rise to m qubits. In order to generate a two-qubit gate we need to implement rotations between the states of two subsystems, such as the ones shown in Figure 3.4(b). As an example we take the U(2) rotation in the tensor product basis of two subsystems, between the states $|11\rangle$ and $|22\rangle$, given by $U(z) = \exp(z|11\rangle\langle 22| - \bar{z}|22\rangle\langle 11|)$, with $z = \theta \exp i\phi$. The corresponding connection components are given by

$$A_\theta = 0, \quad A_\phi = \mathrm{diag}(0, 0, 0, -i\sin^2\theta), \qquad (3.37)$$

written in the basis $\{|00\rangle, |01\rangle, |10\rangle, |11\rangle\}$. A loop C on the (θ, ϕ) plane produces the following holonomy:

$$\Gamma_{\mathbf{A}}(C) = \mathrm{diag}(1, 1, 1, e^{-i\Sigma}), \quad \Sigma = \int_{D(C)} d\theta d\phi \sin 2\theta. \qquad (3.38)$$

For $\Sigma = \pi$ we obtain the controlled-phase CP gate. The ability to produce the CP gate between any qubits together with arbitrary one-qubit rotations leads to universality. This simple model for holonomic quantum computation was first presented in Pachos (2002). A proposal for its physical realisation with trapped ions can be found in Duan *et al.* (2001).

Holonomic quantum computation resembles the adiabatic scheme we have seen in Subsection 3.3.2. The essential ingredient of adiabatic evolution is present in both schemes. The holonomic approach further employs a degenerate space of states and it can have a clear interpretation in terms of quantum gates. In this way it resembles the circuit model. In its turn, topological quantum computation can be considered as holonomic computation where the employed adiabatic evolutions have topological characteristics.

Summary

Quantum computation provides the fascinating perspective of employing the unconventional logic of quantum mechanics to achieve fast information processing. To attain that we need to identify a quantum system with a large Hilbert space where quantum information can be encoded. Quantum information processing is achieved with a set of quantum gates induced by controlling this system with external knobs. When engineering a quantum system as a potential quantum computer one needs to identify a tensor product structure in its Hilbert space that can guarantee the exponential increase in its dimension as a function of its physical size. Moreover, the control of the system needs to be sufficiently rich in order to be able to perform any arbitrary unitary evolution of the Hilbert space. This results in universality, which is our ability to perform any desired quantum algorithm.

There are several ways one could encode and manipulate a quantum system. Here we briefly presented three different ways: one-way quantum computation, adiabatic quantum computation and holonomic quantum computation. All of these are based on different manifestations of quantum phenomena. Fascinating as they might be, quantum computational systems are very fragile. They are naturally coupled to their environment, which can alter or even erase the encoded quantum information through quantum decoherence. To date no physical realisation of quantum computation exists that can offer adequate protection from the environment. Moreover, quantum control procedures need to be refined to be meaningful in manipulating quantum information.

Quantum error correction is an algorithmic method that aims to protect encoded information for sufficiently low error rates. The current bounds on tolerable error rates are formidably low. Moreover, to implement quantum error correction a vast amount of overhead in qubits and quantum gates is required. Topological quantum computers emerged from the idea of trying to address error problems already at the hardware level; it employes some of the most exotic properties of quantum mechanics, namely anyonic statistics, to encode and manipulate information. This approach is naturally fault-tolerant to control errors as well as robust against environmental perturbations. The main elements of topological quantum computation are presented in the next chapter.

Exercises

3.1 Consider (3.9) that describes a system at a given temperature. Demonstrate that at very low temperatures the state of the system corresponds to the ground state. [*Hint*: Use the expansion of the Hamiltonian in terms of its eigenstates.]

3.2 Take a unitary matrix U to be block diagonal of dimension 2^n, for integer $n \geq 2$, with each block being a σ^x matrix. How many single-qubit gates does one need in order to simulate U? If this U was a classical operator how many classical single-bit gates would one need in order to simulate it? [*Hint*: Consider the tensor product structure of n qubits.]

3.3 Consider the one-dimensional cluster state with two qubits. In which basis do we need to measure one of the qubits in order to prepare the second qubit in the state $(|0\rangle + i|1\rangle)/\sqrt{2}$?

3.4 Consider the geometric evolutions of two systems as given in Subsection 3.3.3. Having the composite system initiated in the state $|\psi\rangle = |00\rangle$, which loops in the parametric space and in what order does one need to traverse in order to obtain a maximally entangled state?

Topological quantum computation encodes and manipulates information by exclusively employing anyons. To study the computational power of anyons we plan to look into their fusion and braiding properties in a systematic way. This will allow us to identify a Hilbert space, where quantum information can be encoded fault-tolerantly. We also identify unitary evolutions that serve as logical gates. It is an amazing fact that fundamental properties, such as particle statistics, can be employed to perform quantum computation. As we shall see below, the resilience of these intrinsic particle properties against environmental perturbations is responsible for the fault-tolerance of topological quantum computation.

Anyons are physically realised as quasiparticles in topological systems. Most of the quasiparticle details are not relevant for the description of anyons. This provides an additional resilience of topological quantum computation against errors in the control of the quasiparticles. In particular, the principles of topological quantum computation are independent of the underlying physical system. We therefore do not discuss its properties in this chapter. The abstraction might create a conceptual vacuum as many intrinsic properties of the system might appear to be absent. For example, we shall not be concerned with the trapping and transport of anyons or with geometrical characteristics of their evolutions. In this chapter we treat anyons as classical fundamental particles, with internal quantum degrees of freedom, much like the spin. Moreover, we assume that we have complete control over the topological system, in terms of initial-state preparation and final-state identification. Details of how this can be achieved on concrete physical implementations can be found in later chapters, where explicit topological systems are considered.

This chapter presents the inception of anyonic models. It introduces the necessary steps to consistently define an anyonic theory from basic principles. The first step is to define a finite set of anyonic particles, or species. We identify the vacuum of this set as the trivial particle. Moreover, every particle should have its own antiparticle. These particles are characterised by internal degrees of freedom, which are associated with quantum numbers. Relations between these quantum numbers are obtained by the fusion rules, which dictate what types of species are obtained when combining two particles together. The next step is to verify if the defined model satisfies a set of consistency conditions. These are known as the pentagon and hexagon equations, named after their characteristic geometric configuration. A discrete set of solutions is obtained from these equations that determine the braiding properties of the particles. The fusion and braiding properties are sufficient to obtain concrete models that can be used for topological quantum computation.

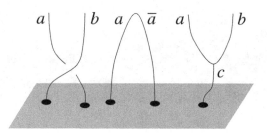

Fig. 4.1 Worldlines of particles that are positioned on a plane with time flowing downwards. An exchange of two particles, a and b, is depicted in terms of braided worldlines. A pair-creation from the vacuum of a particle a and its antiparticle $\bar{a}$ corresponds to two lines initiated at the same position. Fusion of two particles a and b is given by two lines coming together and producing a third particle c.

4.1 Anyons and their properties

We now present the fundamental properties of anyons in a systematic way. It is convenient to keep track of the anyon history by employing their worldlines. In this way we can easily visualise statistical processes and predict the time evolution of anyonic states. Examples of such processes are depicted in Figure 4.1. There, we assume that we can trap and move anyons around the plane leading to worldlines in $2 + 1$ dimensions. Exchanges of two anyons can be described by just braiding their worldlines. We can also depict the pair-creation of anyons from the vacuum as well as the fusion process that occurs when they are brought together, thereby resulting in a new anyon. Since the worldlines represent topological evolutions, no attention needs to be paid to their exact shape. We only need to focus on their global characteristics.

4.1.1 Particle types

Our starting point is to recognise that there can be a variety of different anyonic models. Each model is determined by the statistical properties of its particles. Let us consider such a particular model. To describe it we introduce finitely many different species of particles. When they are realised in a topological system they correspond to quasiparticle excitations that can be distinguished according to their properties with respect to certain physical observables. In the following we shall use the terminologies 'particle', 'anyon' and 'quasiparticle' in an interchangeable way.

Consider a set of particle types

$$1, \ a, \ b, \ c, \ldots, \tag{4.1}$$

where 1 corresponds to the unique vacuum, while $a, b, c, \ldots$ correspond to a finite series of different particle types. The simplest non-trivial model contains one more particle than the vacuum. Every particle, a, needs to have its own antiparticle, $\bar{a}$, which could be itself, so that they can be pair-created from the vacuum. Each particle can be locally distinguished by its topological or anyonic charge, which is a conserved quantum number. For example, the anyonic charge indicates whether a particle corresponds to the vacuum, to a boson or to a fermion. Particles with richer anyonic properties can be similarly identified. The anyonic charge is better described in combination with the rest of the anyonic particles in the system, as we shall see in the following.

4.1.2 Fusion rules of anyons

We now consider the fusion properties of finitely many anyons that belong to a given model. The fusion corresponds to bringing two anyons together and determines how they behave collectively. No interactions need to take place between the anyons. Fusion can be viewed as putting two anyons in a box and identifying the statistical behaviour of the box. For example, fusing two fermions together produces a boson. In general, the fusion rules are written as

$$a \times b = N_{ab}^c c + N_{ab}^d d + \cdots \tag{4.2}$$

These rules indicate the possible outcomes $c, d, \ldots$, listed with the $+$ symbol, that result when anyons a and b are brought together, denoted by the $\times$ symbol. The ordering of a and b is not important, so that

$$a \times b = b \times a. \tag{4.3}$$

When a and b are fused there might be several distinct mechanisms that produce particle c, enumerated by the integers N_{ab}^c.

It is possible to prepare two anyons in a certain way so that they have a unique fusion outcome. For example, two anyons produced from a vacuum pair-creation have the vacuum as their unique fusion channel. So the several possible outcomes on the right-hand side of (4.2) could be understood as different possible preparations of a and b that would result in a certain fusion outcome. Finally, the fusion process can be time-reversed. Consider the case where the fusion of a and b gives a specific fusion outcome c. When time is inverted the same process describes the splitting of anyon c into its constituent particles a and b.

Anyons are systematically characterised by their fusion behaviour. For example, Abelian anyons have only a single fusion channel

$$a \times b = c. \tag{4.4}$$

Their fusion space is one-dimensional. In contrast, non-Abelian anyons always have multiple fusion channels that give rise to higher-dimensional fusion spaces

$$\sum_c N_{ab}^c > 1. \tag{4.5}$$

This simple property is closely related to their statistical behaviour. As we shall see below, non-Abelian statistics is manifested as a non-trivial evolution between the different

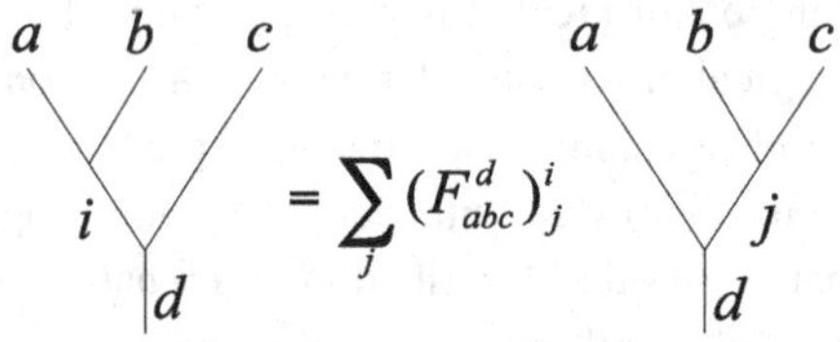

$$\begin{array}{c}
a \quad b \quad c \\
\diagdown\diagdown\diagup \\
i\diagdown\diagup \\
\Big|d
\end{array}
\;=\; \sum_{j} (F^{d}_{abc})^{i}_{j}
\begin{array}{c}
a \quad b \quad c \\
\diagdown\diagdown\diagup \\
\diagdown\diagup j \\
\Big|d
\end{array}$$

Fig. 4.2 When the order of fusion between three anyons, a, b and c, with outcome d is changed then a rotation in the fusion space is performed given by the matrix F^{d}_{abc}. In this diagrammatic equation the index i denotes a certain anyon, while the summation in j ranges through all possible fusion outcomes of b and c.

possible fusion outcomes given in (4.2). Abelian statistics corresponds to the evolution of a unique state by a phase factor.

When we fuse several anyons, we are free to choose the ordering in which the basic fusion processes take place. For example, three anyons, a, b and c, with total fusion channel d can be fused in two different ways. Fusing a and b might have an outcome i that is different from the outcome j of fusing b and c. These are the only two distinctive possible orders in which one can fuse three anyons. Having i and j different is consistent with having a fixed total fusion outcome, d. Explicitly, fusing i with c gives d and fusing j with a gives d as well, as shown in Figure 4.2. This is much like the different ways one can combine several spin-1/2 particles to obtain a given value for the spin of their composite. The matrix F^{d}_{abc} with i, j elements $(F^{d}_{abc})^{i}_{j}$ that relates these two different processes is called the fusion or F matrix and its action is illustrated in Figure 4.2. The dimensionality of this matrix depends on the number of possible in-between outcomes of the fusions.

The choice of fusion order is a degree of freedom in the description of several anyons. Indeed, a sequence of anyons fused in a particular order provides a set of possible in-between fusion outcomes. Another person who has exactly the same set of anyons and decides to fuse them in a different order could obtain a different sequence of in-between fusion outcomes. The F matrix can be employed to systematically translate between these two different sets. Actually, any fusion ordering can be mapped to any other with a sufficient number of F move applications, like the ones depicted in Figure 4.2. Choosing the order in which anyons are fused can be viewed as a choice of basis and the F matrix as a transformation between different bases.

4.1.3 Anyonic Hilbert space

The Hilbert space of anyons is rather unusual. It is the space of states that corresponds to the fusion process. We assign a distinct state to the time evolution of two anyons that fuse to a certain outcome. In this way, states that correspond to different fusion outcomes are automatically orthogonal to each other as we can always distinguish between different anyons. Let us denote the fusion Hilbert space of n anyons by $\mathcal{M}_{(n)}$. Since Abelian anyons have only a single fusion outcome, their fusion Hilbert space is trivial

$$\dim(\mathcal{M}_{\text{Abelian}}) = 1. \tag{4.6}$$

Suppose we consider two non-Abelian anyons a and b with the fusion rule $a \times b = \sum_c N^c_{ab} c$, as in (4.2). In this case we assign the state

$$|a, b \to c; \mu\rangle \tag{4.7}$$

to each possible fusion outcome. The index $\mu = 1, \ldots, N^c_{ab}$ parameterises the possible multiplicity of a certain fusion channel. To simplify notation we restrict ourselves in the following to the case where $N^c_{ab} \leq 1$, so we can drop the index μ.

Let us have a closer look at a variety of fusion processes and their corresponding dimensionality. The hypothetical evolution of a single non-trivial anyon going through fusion with the vacuum and coming out as the vacuum is not permitted. This allows us to assign the zero-dimensional Hilbert space to this evolution

$$\dim(\mathcal{M}_{(1)}) = 0. \tag{4.8}$$

When there is an initial and a final anyon, that due to anyonic charge conservation are necessarily equal to each other, then the Hilbert space is one-dimensional

$$\dim(\mathcal{M}_{(2)}) = 1. \tag{4.9}$$

The Hilbert space of two initial non-Abelian anyons a and b with a fusion outcome c with multiplicity N^c_{ab} gives rise to equally many states. The Hilbert space of three anyons related by the fusion process is therefore given by

$$\dim(\mathcal{M}_{(3)}) = N^c_{ab}. \tag{4.10}$$

Consider now three initial anyons a, b and c that fuse to d. To evaluate the dimension of their Hilbert space we need to count all possible in-between outcomes from pairwise fusions. To be explicit we can initially fuse a and b and then fuse the outcome i of this fusion with c in order to obtain d, as illustrated in Figure 4.3(a). For each i we might write the state of this fusion process as

$$|i\rangle = |a, b \to i\rangle |i, c \to d\rangle, \tag{4.11}$$

where the tensor product symbol between the states of the two different fusion processes has been omitted. This state can be written as $|i\rangle$ when the fixed anyons a, b, c and d are implicitly assumed. If there is more than one possible outcome then the corresponding states, $|i\rangle$, can comprise a basis of a higher-dimensional Hilbert space denoted $\mathcal{M}_{(4)}$. Alternatively, one could consider fusing b and c and their outcome j with a to obtain d with

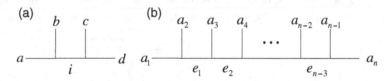

Fig. 4.3 (a) Basis states $|i\rangle$ for the fusion space of four ordered anyons a, b, c and d. (b) Basis states $|e_1 e_2, \ldots, e_{n-3}\rangle$ of n ordered anyons a_i with $i = 1, \ldots, n$.

corresponding basis states $|b, c \to j\rangle |j, a \to d\rangle$. Changing between these two different fusion states corresponds to the F move we described in Figure 4.2. The change of basis in the Hilbert space $\mathcal{M}_{(4)}$ of the four anyons is given by

$$|a, b \to i\rangle |i, c \to d\rangle = \sum_j (F_{abc}^d)_j^i |b, c \to j\rangle |a, j \to d\rangle \qquad (4.12)$$

or simply $|i\rangle = \sum_j (F_{abc}^d)_j^i |j\rangle$. If we consider more initial anyons we have to specify how we order their fusions if we want to uniquely determine the basis states of their Hilbert space $\mathcal{M}_{(n)}$. For the ordering of n anyons a_i with $i = 1, \ldots, n$, depicted in Figure 4.3(b), we have the states

$$|\mathbf{e}\rangle = |e_1, e_2, \ldots, e_{n-3}\rangle = |a_1, a_2 \to e_1\rangle |e_1, a_3 \to e_2\rangle \ldots |e_{n-3}, a_{n-1} \to a_n\rangle. \quad (4.13)$$

By a simple counting argument we can see that the number of different fusion possibilities is given by

$$\dim(\mathcal{M}_{(n)}) = \sum_{e_1 \ldots e_{n-3}} N_{a_1 a_2}^{e_1} \ldots N_{e_{n-3} a_{n-1}}^{a_n}. \qquad (4.14)$$

A more intuitive expression for $\dim(\mathcal{M}_{(n)})$ can be given in terms of the quantum dimension d_i of anyon i. Quantum dimension is a fancy name that refers to the dimension of the Hilbert space associated with an anyon. Starting from the fusion rules $a \times b = \sum_c N_{ab}^c c$ one can define the quantum dimension to satisfy the following relation:

$$d_a d_b = \sum_c N_{ab}^c d_c. \qquad (4.15)$$

Abelian anyons, such as the vacuum, always have $d_i = 1$, while non-Abelian anyons necessarily have $d_i > 1$. It is worth noting that the quantum dimension does not need to be an integer. Consider now the set of n anyons, shown in Figure 4.3(b), where all a_i are identical to a. The quantum dimension characterises how fast the dimension of the Hilbert space grows when one additional a particle is inserted, i.e.,

$$\dim(\mathcal{M}_{(n)}) \propto d_a^n, \qquad (4.16)$$

where we assume that n is large (see Exercise 4.2). The dimension of $\mathcal{M}_{(n)}$ is always an integer as it enumerates different fusion outcomes, while d_a^n does not need to be an integer. Relation (4.16) therefore gives the proper behaviour only for large n. The important fact is that the fusion Hilbert space increases exponentially fast with the number of anyons n. Nevertheless, $\dim(\mathcal{M}_{(n)})$ is not necessarily the product of the dimensions of particular subsystems. Finally, we define the total quantum dimension of a topological model by

$$\mathcal{D} = \sqrt{\sum_i d_i^2}, \qquad (4.17)$$

where the summation runs through all the anyonic species of the model. The quantity $\mathcal{D}$ can be defined for any topological model.

Before moving further let us interpret these rather obscure fusion states in terms of more conventional means. After all, when the topological model is physically realised, the fusion states have to correspond to certain quantum states of the constituent particles. We expect that the states of the microscopic system which correspond to different fusion outcomes are pairwise orthogonal. On the other hand, microscopic states that produce the same fusion outcome are considered as equivalent. This is manifested as an indistinguishability of the microscopic states in terms of their topological properties. The information on the fusion outcome is not a local property as it is encoded in the system in a non-local way. For example, consider two quasiparticles a and b prepared in a given fusion channel c. Their fusion state $|a, b \rightarrow c\rangle$ corresponds to a concrete state of the underlying microscopic system. When this state evolves adiabatically in order to fuse anyonic quasiparticles, the state that corresponds to quasiparticle c results. All the states of the constituent particles along this time evolution that describe different positions of the a and b quasiparticles are equivalent since they correspond to the same fusion state. As a conclusion the fusion states correspond, in general, to a whole family of states of the microscopic system.

4.1.4 Exchange properties of anyons

Statistics is manifested in the evolution of the wave function of two particles when they are exchanged. In two spatial dimensions particles are allowed to exhibit any arbitrary statistical evolution. To systematically assign statistical evolutions consider the effect of exchanging two anyons, a and b, when their fusion channel is fixed, i.e., $a \times b \rightarrow c$, as shown in Figure 4.4. This exchange can be viewed as a half twist of the c particle. Hence, the exchange evolution R_{ab}^c of the fusion state $|a, b \rightarrow c\rangle$ should simply be a phase factor as it corresponds to the rotation of a single particle. We can build a matrix R_{ab} by ordering the phases for all possible fusion outcomes c of a and b on its diagonal. This exchange matrix will be referred to in the following as the R matrix.

The superposition of multiple fusion outcomes in the braiding process can result in an exchange operator B, which is a non-diagonal unitary matrix. To demonstrate this, we consider the effect of exchanging a and b when these two anyons do not have a direct fusion channel. Then the F moves can be employed to change their fusion order until the exchange is acting on anyons with a certain fusion channel. In Figure 4.5 we derive diagrammatically the relation

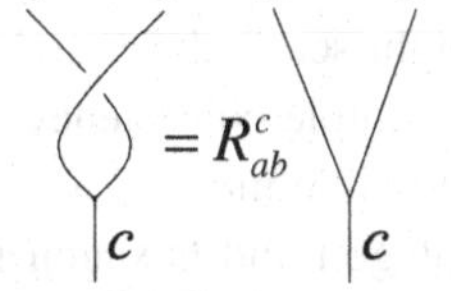

Fig. 4.4   The clockwise exchange of anyons a and b with fusion outcome c gives the phase R_{ab}^c.

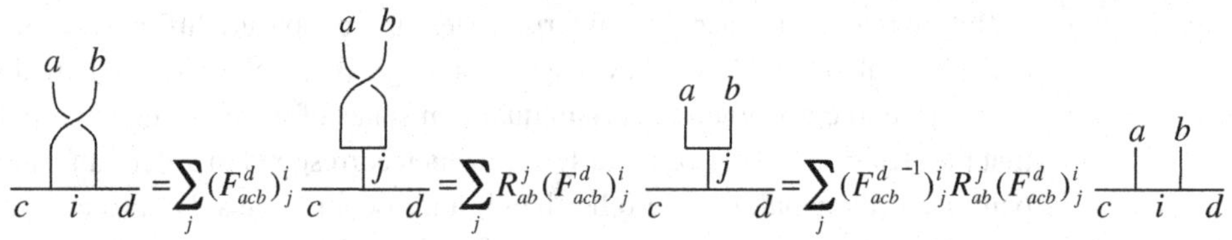

$$\underset{c}{\overset{a\ b}{\bigvee}}_{i\ d} = \sum_j (F^d_{acb})^i_j \underset{c}{\overset{a\ b}{\bigvee}}_{j\ d} = \sum_j R^j_{ab}(F^d_{acb})^i_j \underset{c}{\overset{a\ b}{\big|\big|}}_{j\ d} = \sum_j (F^{d\ -1}_{acb})^i_j R^j_{ab}(F^d_{acb})^i_j \underset{c}{\overset{a\ b}{\big|\big|}}_{i\ d}$$

Fig. 4.5 If anyons a and b do not have a direct fusion channel then their exchange can be defined in terms of the F moves that rearrange the order of fusion. Here we depict such a series of operations that gives rise to the braiding unitary $B = F^{-1}RF$.

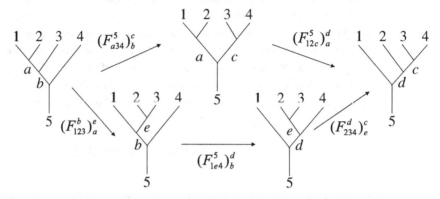

Fig. 4.6 Geometric interpretation of the pentagon identity. We start with the canonical fusion of four anyons with a fixed total outcome given by the leftmost diagram. Two different sequences of F moves applied on the initial fusion diagram arrive at the same final diagram. It is taken as an axiom that these two sequences are identical.

$$B_{ab} = F^{d\ -1}_{acb} R_{ab} F^d_{acb}. \tag{4.18}$$

The braiding matrix B_{ab} depends implicitly on anyons c and d. Notably, this unitary matrix can be non-diagonal due to the F transformation. The B unitary corresponds to irreducible representations of the braid group. The general properties of the braid group are analysed in Chapter 8.

4.1.5 Pentagon and hexagon identities

Arbitrary as they might seem, the F and R matrices that accompany a given set of fusion rules have to satisfy simple consistency equations. These conditions dramatically restrict the multiplicity of possible models, which satisfy the same fusion rules, to finitely many. They are called pentagon and hexagon identities (Turaev, 1994) due to their geometric interpretation and they are the subject of study of topological quantum field theory (Witten, 1989).

Let us consider Figure 4.6, where the fusion process of four anyons, 1, 2, 3 and 4, is depicted. Consider the leftmost diagram with a certain fusion ordering. We assign specific in-between fusion outcomes, a and b, that have a fixed total fusion channel, 5. By employing the two F moves depicted in the upper path it is possible to completely reverse the fusion ordering and transform the fusion diagram to the rightmost one. However, it is also possible to connect these two diagrams by following a completely different path that includes three F moves. This is depicted in Figure 4.6 as the lower path. It is an axiom that these two processes should be equivalent. Stated differently, if there is a unique interpretation of fusion states by the fusion diagrams then distinct transformations with F moves that connect the leftmost and rightmost diagrams ought to be identical. Imposing this axiom gives rise to the pentagon identity

$$(F^5_{12c})^d_a(F^5_{a34})^c_b = \sum_e (F^d_{234})^c_e(F^5_{1e4})^d_b(F^b_{123})^e_a. \tag{4.19}$$

This equation provides a relation between the matrix elements of all possible F matrices of the model. The e summation is over all possible particle types that we can have in the fusion diagrams shown in Figure 4.6.

An independent set of identities can be obtained by employing in addition the braiding processes. Consider three anyons, 1, 2 and 3, that fuse to 4 through the fusion channel a, as shown in the leftmost diagram of Figure 4.7. By alternating applications of F and R moves it is possible to interchange the fusion order of the initial anyons in two distinct ways. Demanding again that these two distinct processes correspond to the same overall procedure gives rise to the hexagon identity

$$\sum_b (F^4_{231})^c_b R^4_{1b}(F^4_{123})^b_a = R^c_{13}(F^4_{213})^c_a R^a_{12}. \tag{4.20}$$

When instead counterclockwise exchange operations R^{-1} are employed, an equivalent set of equations is obtained (Bonderson, 2007).

Finally note that the pentagon and hexagon identities become trivial for Abelian models, whose statistical phase can be arbitrary. Consistent non-Abelian anyonic models are completely determined by the pentagon and hexagon identities without the need for further conditions (MacLane, 1998). For a given number of anyon types with fixed fusion rules,

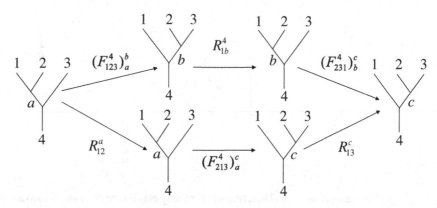

Fig. 4.7   The hexagon identity relates two distinct fusion processes of three anyons with a fixed total fusion outcome by a sequence of fusion rearrangements and braiding operations.

the solutions of these two polynomial equations give a discrete possibly empty set of F and R matrices. This resembles the discrete character of the solutions of quadratic equations. The discreteness in the F and R solutions, known as the Ocneanu rigidity (Kitaev, 2006) is in agreement with the discrete nature of topological models. Hence, topological models are not continuously connected with each other, which provides much of their resilience against erroneous perturbations.

4.1.6 Spin and statistics

It is well known that bosons have integer spin (e.g., 0) and fermions half-integer spin (e.g., 1/2). When a spin-0 particle is rotated around itself its wave function is not changed, while when a spin-1/2 particle is rotated by 2π then the fermionic wave function acquires a minus sign (Rauch *et al.*, 1975). This is in agreement with the exchange statistics of these particles. The tight relation between spin and statistics (Finkelstein and Rubinstein, 1968) also governs the behaviour of anyons. As the statistics of anyons is neither bosonic nor fermionic, the spin of anyons can take any value different from 0 or 1/2. Up to now the worldlines of anyons allowed us to keep track of their braiding history. To keep track of their self-rotations we now extend the worldlines to 'worldribbons'. This allows us to establish the connection between spin and statistics.

Consider two anyons a and b with a given fusion channel c that are exchanged k times in a clockwise fashion. Particle exchanges cannot change the fusion outcome of a and b, but they can generate phase factors, R_{ab}^c, as we have seen previously. Suppose that the quantum mechanical evolution, associated with the exchange process, remains invariant under continuous deformations of the worldribbons, due to its topological character. Then it is possible to continuously transform the k clockwise exchanges of a and b to k twists of the ribbons by an angle π, clockwise for ribbon c and counterclockwise for ribbons a and b. This is depicted in Figure 4.8. The spin-statistics theorem dictates that the amplitudes of these two processes have to be equal (Finkelstein and Rubinstein, 1968). Let us assign spins, s_a, s_b and s_c to anyons a, b and c, respectively. Clockwise rotating a spin s particle

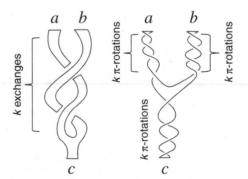

Fig. 4.8 Interpreting anyons with ribbons facilitates accounting for twists and exchanges. Clockwise exchanging k times anyons a and b can be continuously deformed to k clockwise π rotations for anyon c and k counterclockwise π rotations for both anyons a and b. These two configurations are topologically equivalent.

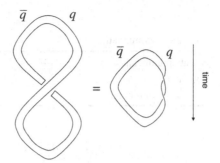

Fig. 4.9   An anyon q and an anti-anyon $\bar{q}$ denoted with their worldribbons are pair-created, exchanged and then fused to the vacuum. The exchange process can be continuously deformed to rotating one of the anyons around itself by 2π. This equivalence can be nicely verified with a belt.

by angle ϕ generates the phase factor $e^{-i\phi s}$ in front of its wave function. As the amplitude of the exchange and the twisting processes have to be the same, the twists of the particles have to generate the appropriate spin phase factors to compensate for the statistical ones. Applying this to the process of Figure 4.8, we obtain the spin-statistics theorem given in the form (Bais *et al.*, 1992)

$$(R_{ab}^c)^k = e^{i\pi k s_a} e^{i\pi k s_b} e^{-i\pi k s_c}. \tag{4.21}$$

As an example, we consider the $k = 1$ case. We restrict ourselves to anyons $a = q$ and $b = \bar{q}$ that are particles and antiparticles of each other with the vacuum being their fusion channel, $c = 1$. The corresponding spins are given by $s = s_q = s_{\bar{q}}$ and $s_1 = 0$. As these anyons can be generated from the vacuum and fused to it, their evolution corresponds to a worldribbon that forms a closed loop. In Figure 4.9 we show the schematic equivalence between the process of exchanging q and $\bar{q}$ and rotating only one of them by 2π. The first evolution gives rise to a statistical phase $R_{q\bar{q}}^1$ and the second to a spin phase $e^{i2\pi s}$, where s is the spin of the q and $\bar{q}$ anyons. Hence,

$$R_{q\bar{q}}^1 = e^{i2\pi s}, \tag{4.22}$$

which is exactly the spin-statistics relation. As Abelian anyons can have arbitrary values of statistic phases $e^{i\varphi}$, their corresponding spin s can take arbitrary values as well.

4.2   Anyonic quantum computation

In the previous sections we identified the Hilbert space of non-Abelian anyons and analysed the manipulations that lead to unitary evolutions of this space. We are now ready to see how to employ anyons to perform quantum computation. For that we need to implement several operations on anyons to eventually achieve the desired quantum state manipulations. Our steps follow the circuit quantum computation model. This model requires initialisation of the physical system in a well-determined quantum state, application of quantum gates

Table 4.1 Anyonic quantum computation		
Quantum computation		Anyonic manipulation
State initialisation	$\rightarrow$	Create and arrange anyons
Quantum gates	$\rightarrow$	Braid anyons
State measurement	$\rightarrow$	Detect anyonic charge

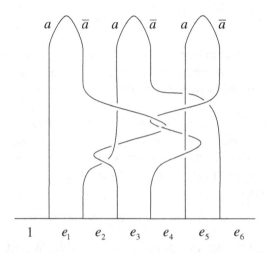

Fig. 4.10 A possible configuration of topological quantum computation. Initially, pairs of anyons, $a, \bar{a}$ are created from the vacuum. Braiding operations between them unitarily evolve their fusion state. Finally, fusing the anyons together gives a set of outcomes $e_i, i = 1, \ldots$, which encodes the result of the computation.

and measurement of the final state. To implement these steps we seek to initially create and arrange anyons, braid them together and eventually determine their anyonic charge, as summarised in Table 4.1.

4.2.1 Anyonic setting

A possible configuration and manipulation of anyons that can result in quantum computation is shown in Figure 4.10. We start with a set of anyons that are prepared in a well-defined fusion state. For example, this is possible by creating pairs of non-Abelian anyons a and $\bar{a}$ from the vacuum. The fusion state of these anyons is well known. It belongs to a Hilbert space that increases exponentially with the number, n, of anyonic pairs, $\dim(\mathcal{M}_{(n)}) \propto d_a^n$. As d_a is not always an integer, the fusion space $\mathcal{M}_{(n)}$ does not necessarily admit a tensor product structure. Nevertheless, this Hilbert space admits a subspace with qubit tensor product structure in which quantum information can be encoded in the usual way. Its dimension increases exponentially as a function of n. Hence, non-Abelian anyons are an efficient medium for storing quantum information.

Having identified the logical encoding space we now consider the gates that evolve it. Logical gates can be performed by braiding the anyons, thus evolving their fusion state by the R matrix, as shown in Figure 4.10. This operation does not affect either the type of anyons or their local degrees of freedom, but can have a non-trivial effect on the states of the fusion space $\mathcal{M}$, as we have shown in Subsection 4.1.4. In combination with the F matrices one can evolve the encoded information in a non-trivial way. Ideally, we want to be able to perform any arbitrary algorithm out of braiding anyons. Assume that the F and the R matrices span a dense set of unitaries acting on the qubits, in the sense described in Subsection 3.2.1. Then the corresponding anyonic model supports universal quantum computation implemented just by braiding anyons (Freedman *et al.*, 2002a, b). For these models it has been shown (Burrello *et al.*, 2010; Simon *et al.*, 2006) that by weaving a single anyon among many static ones it is possible to perform a universal set of gates between arbitrary qubits. Then one can employ these gates to implement quantum computation following standard quantum algorithms.

At the end of the computation we want to measure the processed information, which is encoded in the final fusion state of the anyons. This can be achieved by fusing the anyons in a series and retrieving the fusion outcomes e_i. The example given in Figure 4.10 illustrates this process at the end of the anyonic evolution. As the fusion state of the anyons can be a superposition of many different basis states $|e_1, e_2, \ldots\rangle$, the measurement of the final fusion state provides, in general, a probability distribution. This step constitutes the final read out of the computation. The braiding algorithm can be adapted to different choices of initial states of anyons and to different fusion procedures.

4.2.2 Stability of anyonic computation

Let us now have a closer look at the stability features of topological quantum computation. Initially, note that the fusion space evolution induced by anyon braiding does not depend on the details of the paths spanned by the anyons, only on their topology. The experimental control of the system inherits this resilient characteristic. Hence, an experimentalist implementing topological quantum computation does not need to be very careful in spanning these paths as long as their global characteristics are realised.

If anyons were elementary particles then they would be robust up to high energy scales. Hence, information encoded with the anyons would be resilient and we could straightforwardly perform error-free quantum computation. In reality, anyons are realised as effective particles of topological models. Thus, we need to consider the stability of these models against environmental errors. What protects the logical information encoded in these systems is the non-local encoding and the presence of a finite energy gap. Indeed, when anyons are kept far apart the information encoded in the fusion space is not accessible by local operations. Hence, environmental errors, acting as local perturbations to the Hamiltonian cannot alter the fusion states (Bravyi *et al.*, 2010). This is the fault-tolerant characteristic of anyons that makes them a favourable medium for performing quantum computation. Nevertheless, probabilistic errors on the system (e.g., due to a finite temperature) do affect the encoded space (Bravyi *et al.*, 2009). It is an important open problem to find a method that efficiently overcomes probabilistic errors with a two-dimensional system. First important steps are taken in Chesi *et al.* (2010) and Hamma *et al.* (2009).

Finally, we should emphasise that implementing universal computation solely by topological means is not the only available option. One might envision combining topological procedures with other known computational methods to optimise their resilience and efficiency. For example, quantum information can be stored in the fusion channels of anyons and thus become protected from errors compared to other quantum memory schemes. Subsequently, one might like to avoid transporting anyons in order to perform logical gates and instead perform them in a dynamical, non-topological way. A scheme has already been proposed that employs measurements of anyons in order to evolve their state, similarly to one-way quantum computation (Bonderson *et al.*, 2008; 2009). Moreover, for some models, the braiding and recombining operations might not be enough to span a universal set of gates while they still provide an efficient anyonic quantum memory. Supplementing these operations with non-topological evolutions can lead to universal quantum computational models (Bravyi, 2006; Das Sarma *et al.*, 2005).

4.3   Example I: Ising anyons

To illustrate the properties of anyonic models we now consider the example of the Ising anyons. The importance of this non-Abelian anyonic model stems from the fact that it is the most promising model for experimental realisation. As we shall see in Chapter 6, Ising anyons describe the statistical properties of Majorana fermions. The latter are currently under intense experimental investigation in the arena of fractional quantum Hall samples (Miller *et al.*, 2007), topological insulators (Fu *et al.*, 2007) and *p*-wave superconductors (Read and Green, 2000).

4.3.1   The model and its properties

The particle types of the Ising anyon model are the vacuum, 1, the non-Abelian anyon, σ, and the fermion, ψ. In this model the fusion rules are given by

$$\sigma \times \sigma = 1 + \psi, \quad \sigma \times \psi = \sigma, \quad \psi \times \psi = 1, \tag{4.23}$$

with 1 fusing trivially with the rest of the particles (i.e., $\sigma \times 1 = \sigma$ and $\psi \times 1 = \psi$). The first fusion rule of (4.23) signifies that if we bring two σ anyons together they might annihilate (i.e., σ can be its own antiparticle) or they might give rise to the fermion ψ. Hence, the fusion of two σ's has two possible fusion outcomes represented by the states $|\sigma, \sigma \rightarrow 1\rangle$ and $|\sigma, \sigma \rightarrow \psi\rangle$. The second rule indicates that fusing a ψ with a σ gives back a σ. In a sense, this rule states that a ψ can be absorbed by a σ without changing its anyonic charge. The third fusion rule states that when two fermions are brought together they are fused to the vacuum. Only the parity of the total number of fermions can be detected, since the composite of two ψ's is condensed to the vacuum.

Let us now give the explicit forms of the F and R matrices for the Ising model. We postpone their derivation to the next subsection. The F matrix is given by

$$F^\sigma_{\sigma\sigma} = \frac{1}{\sqrt{2}} \begin{pmatrix} 1 & 1 \\ 1 & -1 \end{pmatrix} \tag{4.24}$$

in the $|\sigma,\sigma \rightarrow 1\rangle$ and $|\sigma,\sigma \rightarrow \psi\rangle$ basis. It corresponds to the rearrangement of the fusion order of three σ anyons when their total fusion channel is a σ. The F matrix dictates that the in-between fusion outcomes, being the vacuum or the fermion, can be non-trivially transformed by changing the fusion order of the anyons. In the case of two σ anyons the components of the R matrix are $R^1_{\sigma\sigma} = e^{-i\pi/8}$ and $R^\psi_{\sigma\sigma} = e^{i3\pi/8}$, giving the matrix

$$R_{\sigma\sigma} = e^{-i\pi/8} \begin{pmatrix} 1 & 0 \\ 0 & i \end{pmatrix}. \tag{4.25}$$

This implies that a ψ fusion channel acquires an additional $\pi/2$ phase compared to the vacuum during a π rotation due to the spin-$1/2$ nature of the fermion.

Let us now consider some implications of the braiding and fusion properties of the Ising anyons. Assume that one creates two pairs of anyons (σ,σ) from the vacuum, as illustrated in Figure 4.11(a). The state of the two pairs is then given by $|\sigma,\sigma \rightarrow 1\rangle|\sigma,\sigma \rightarrow 1\rangle$. The braiding evolution is described by the two-dimensional matrix

$$B = F^{-1}R^2 F = e^{-i\pi/4} \begin{pmatrix} 0 & 1 \\ 1 & 0 \end{pmatrix} \tag{4.26}$$

that rotates the fusion states of each pair from $|\sigma,\sigma \rightarrow 1\rangle$ to $|\sigma,\sigma \rightarrow \psi\rangle$ up to an overall phase factor. The braiding, hence, changes the internal state of the anyons in a non-trivial way. The resulting ψ's can be further fused to the vacuum that we had started with, without violating the conservation of the total anyonic charge. Similarly, Figure 4.11(b) shows the

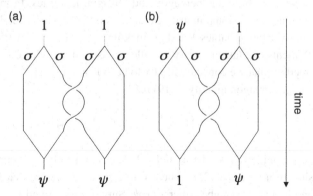

Fig. 4.11 Worldlines of Ising anyons, σ, where time is running downwards. (a) Two pairs of σ's are generated from the vacuum 1. Then an anyon from one pair is circulated around an anyon from the other pair. Finally, the anyons are pairwise fused producing fermionic outcomes. This signals a non-trivial evolution of the fusion states due to braiding. (b) A similar evolution, where two pairs of σ's are created from a fermion ψ and the vacuum 1, respectively. The braiding causes the teleportation of the fermion from one pair to the other.

generation of one pair of Ising anyons from a fermion and another one from the vacuum. The braiding process causes the fermion to be teleported from one pair to the other, even though the anyons have not been in contact with each other at any time.

Let us now describe how one could employ Ising anyons for topological quantum computation. First, we encode a qubit in a set of four σ anyons. Logical states are encoded in the different in-between fusion outcomes of four anyons, i.e. $|0_L\rangle = |\sigma, \sigma \to 1\rangle$ and $|1_L\rangle = |\sigma, \sigma \to \psi\rangle$. To encode n qubits we can employ $4n$ anyons. Logical operations between the qubits can be performed by braiding Ising anyons and changing their fusion order. As we have seen, this results in the F and R matrices given in (4.24) and (4.25), respectively. It is known that these two unitary evolutions cannot support universal quantum computation as F and R do not span the whole SU(2) group. They are restricted to the Clifford subgroup of SU(2). This model can be made universal by the addition of a phase gate which can be implemented by dynamical operations (Bravyi, 2006; Das Sarma *et al.*, 2005).

4.3.2 *F* and *R* matrices

We now explicitly calculate the F and R matrices for the Ising model. The starting point is the set of particles 1, σ and ψ and their fusion rules (4.23). Having a closer look at these rules we find that the only non-zero coefficients are $N^1_{\sigma\sigma} = 1$, $N^\psi_{\sigma\sigma} = 1$, $N^\sigma_{1\sigma} = 1$ and $N^\sigma_{\psi\sigma} = 1$. By substituting $a_1 = a_2 = a_3 = a_4 = \sigma$ into equation (4.14) and having the summation running over $e_1 = 1, \psi$ we find that $\dim(\mathcal{M}_{(4)}) = 2$, which can encode one qubit. Moreover, the quantum dimension of the vacuum is $d_1 = 1$, and ψ has quantum dimension $d_\psi = 1$ as well. So for σ we have $d_\sigma^2 = d_1 + d_\psi$, which implies $d_\sigma = \sqrt{2}$. Thus, the total quantum dimension of the Ising model is $\mathcal{D} = 2$.

Next we solve the pentagon and hexagon identities. From Figure 4.2 we see that F^4_{123} is a one-dimensional matrix, except when the anyons 1, 2, 3 and 4 are all σ. Then i and j run over the variables 1 and ψ, making $F^\sigma_{\sigma\sigma\sigma}$ a 2×2 matrix. All the one-dimensional F elements can take arbitrary complex phase values. This corresponds to a gauge degree of freedom that we conveniently fix to be $+1$ or -1.

The pentagon identity (4.19) reads

$$(F^5_{12c})^d_a (F^f_{a34})^c_b = \sum_e (F^d_{234})^c_e (F^f_{1e4})^d_b (F^b_{123})^e_a. \tag{4.27}$$

Let us initially take the particles 1, 2, 3 and 4 to be σ anyons and 5 to be the vacuum, as shown in Figure 4.12(a). Then from Figure 4.6 we see that b and d need to be σ, while a and c of (4.27) can be either 1 or ψ. Suppose $a = 1$ and $c = 1$. Then the pentagon identity becomes

$$(F^1_{\sigma\sigma1})^\sigma_1 (F^1_{1\sigma\sigma})^1_\sigma = \sum_{e=1,\psi} (F^\sigma_{\sigma\sigma\sigma})^1_e (F^1_{\sigma e\sigma})^\sigma_\sigma (F^\sigma_{\sigma\sigma\sigma})^e_1. \tag{4.28}$$

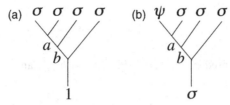

Fig. 4.12 Two initial configurations for the pentagon identity.

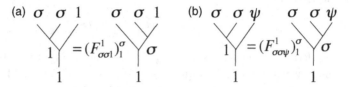

Fig. 4.13 Two trivial F moves. For the (a) configuration we choose $(F^1_{\sigma\sigma1})^\sigma_1 = 1$. The (b) configuration corresponds to an impossible fusion process, so $(F^1_{\sigma\sigma\psi})^\sigma_1 = 0$.

The F move $(F^1_{\sigma\sigma1})^\sigma_1$ corresponds to a trivial rearranging of anyons, as shown in Figure 4.13(a). So we set $(F^1_{\sigma\sigma1})^\sigma_1 = 1$. Hence, we obtain the equation

$$1 = (F^\sigma_{\sigma\sigma\sigma})^1_1{}^2 + (F^\sigma_{\sigma\sigma\sigma})^1_\psi (F^\sigma_{\sigma\sigma\sigma})^\psi_1. \tag{4.29}$$

Let us now take $a = 1$ and $c = \psi$. Then the pentagon identity becomes

$$(F^1_{\sigma\sigma\psi})^\sigma_1 (F^1_{1\sigma\sigma})^\psi_\sigma = \sum_{e=1,\psi} (F^\sigma_{\sigma\sigma\sigma})^\psi_e (F^1_{\sigma e\sigma})^\sigma_\sigma (F^\sigma_{\sigma\sigma\sigma})^e_1, \tag{4.30}$$

which implies

$$(F^\sigma_{\sigma\sigma\sigma})^\psi_\psi = -(F^\sigma_{\sigma\sigma\sigma})^1_1. \tag{4.31}$$

Above we have used the condition $(F^1_{\sigma\sigma\psi})^\sigma_1 = 0$ as the corresponding process, shown in Figure 4.13(b), is forbidden. When $a = \psi$ and $c = 1$ we obtain the same condition as (4.29). Finally, when $a = c = \psi$ we have

$$1 = (F^\sigma_{\sigma\sigma\sigma})^\psi_1 (F^\sigma_{\sigma\sigma\sigma})^1_\psi + (F^\sigma_{\sigma\sigma\sigma})^{\psi\,2}_\psi. \tag{4.32}$$

Let us now take particle 1 to be ψ and particles 2, 3, 4 and 5 to be σ anyons, as shown in Figure 4.12(b). The only possibility is to have $a = d = \sigma$, while b and c can be either 1 or ψ. The pentagon equation, for $b = c = 1$, now becomes

$$(F^\sigma_{\psi\sigma1})^\sigma_\sigma (F^\sigma_{\sigma\sigma\sigma})^1_1 = \sum_{e=1,\psi} (F^\sigma_{\sigma\sigma\sigma})^1_e (F^\sigma_{\psi e\sigma})^\sigma_1 (F^1_{\psi\sigma\sigma})^e_\sigma, \tag{4.33}$$

which implies

$$(F^{\sigma}_{\sigma\sigma})^1_1 = (F^{\sigma}_{\sigma\sigma})^1_{\psi}. \tag{4.34}$$

Here we used $(F^{\sigma}_{\psi\sigma 1})^{\sigma}_{\sigma} = 1$. Finally, for $b = 1$ and $c = \psi$ we obtain

$$(F^{\sigma}_{\psi\sigma\psi})^{\sigma}_{\sigma}(F^{\sigma}_{\sigma\sigma})^{\psi}_1 = \sum_{e=1,\psi}(F^{\sigma}_{\sigma\sigma})^{\psi}_e(F^{\sigma}_{\psi e\sigma})^{\sigma}_1(F^1_{\psi\sigma\sigma})^e_{\sigma}, \tag{4.35}$$

which implies

$$(F^{\sigma}_{\sigma\sigma})^{\psi}_1 = (F^{\sigma}_{\sigma\sigma})^{\psi}_{\psi}. \tag{4.36}$$

To derive this we have set $(F^{\sigma}_{\psi\sigma\psi})^{\sigma}_{\sigma} = -1$ as setting it to $+1$ would have given a non-unitary matrix for $F^{\sigma}_{\sigma\sigma}$. Equations (4.29), (4.31), (4.32), (4.34), (4.36) can now be solved to find that the matrix $F^{\sigma}_{\sigma\sigma}$ has the following elements:

$$(F^{\sigma}_{\sigma\sigma})^{\psi}_{\psi} = -(F^{\sigma}_{\sigma\sigma})^1_1, \quad (F^{\sigma}_{\sigma\sigma})^{\psi}_1 = (F^{\sigma}_{\sigma\sigma})^{\psi}_{\psi},$$

$$(F^{\sigma}_{\sigma\sigma})^1_1 = (F^{\sigma}_{\sigma\sigma})^1_{\psi}, \quad (F^{\sigma}_{\sigma\sigma})^{\psi}_1 = \pm\frac{1}{\sqrt{2}}. \tag{4.37}$$

Reconstructing the F matrix from its components, we obtain

$$F^{\sigma}_{\sigma\sigma} = \pm\frac{1}{\sqrt{2}}\begin{pmatrix} 1 & 1 \\ 1 & -1 \end{pmatrix}. \tag{4.38}$$

Hence, the pentagon equation determines the F matrix. The choice of $\pm$ sign is called the Frobenius–Schur indicator (Rowell *et al.*, 2009).

Let us now turn to the hexagon identity (4.20) given by

$$\sum_b (F^4_{231})^c_b R^4_{1b}(F^4_{123})^b_a = R^c_{13}(F^4_{213})^c_a R^a_{12}. \tag{4.39}$$

In particular, we take particles 1, 2, 3 and 4 to be all σ anyons and consider the four possibilities with a and c being either 1 or ψ. One can easily see that for $a = c = 1$ we have

$$\sum_{b=1,\psi}(F^{\sigma}_{\sigma\sigma})^1_b R^{\sigma}_{\sigma b}(F^{\sigma}_{\sigma\sigma})^b_1 = R^1_{\sigma\sigma}(F^{\sigma}_{\sigma\sigma})^1_1 R^1_{\sigma\sigma}, \tag{4.40}$$

which implies

$$\frac{1}{2}(R^{\sigma}_{\sigma 1} + R^{\sigma}_{\sigma\psi}) = \frac{1}{\sqrt{2}}R^1_{\sigma\sigma}{}^2. \tag{4.41}$$

Similarly, for $a = 1$ and $c = \psi$ we obtain

$$\frac{1}{2}(R^{\sigma}_{\sigma 1} - R^{\sigma}_{\sigma\psi}) = \frac{1}{\sqrt{2}}R^{\psi}_{\sigma\sigma}R^1_{\sigma\sigma}, \tag{4.42}$$

for $a = \psi$ and $c = 1$ we obtain

$$\frac{1}{2}(R^\sigma_{\sigma 1} - R^\sigma_{\sigma \psi}) = \frac{1}{\sqrt{2}} R^1_{\sigma\sigma} R^\psi_{\sigma\sigma} \qquad (4.43)$$

and finally, for $a = c = \psi$ we obtain

$$\frac{1}{2}(R^\sigma_{\sigma 1} + R^\sigma_{\sigma \psi}) = -\frac{1}{\sqrt{2}} R^\psi_{\sigma\sigma}{}^2. \qquad (4.44)$$

Combining (4.41) and (4.44) hence implies

$$R^1_{\sigma\sigma} = \pm i R^\psi_{\sigma\sigma}, \qquad (4.45)$$

while adding (4.42) and (4.44) together gives

$$R^\psi_{\sigma\sigma} = \pm e^{-\frac{3\pi}{8}i}. \qquad (4.46)$$

Similarly, we find that for both choices of sign we have the same solution $R^\sigma_{\sigma 1} = 1$ and $R^\sigma_{\sigma \psi} = i$. Note that there is a discrete multiplicity of solutions in equations (4.38), (4.45) and (4.46) corresponding to the combinations of different signs. Hence, the hexagon equation determines the R matrix when the F matrix is known.

4.4 Example II: Fibonacci anyons

In this final section we present probably the most celebrated non-Abelian anyonic model: the Fibonacci anyons. Its popularity is not only due to its simplicity and richness in structure, which supports universal quantum computation, but also to its connection to the Fibonacci series. In this model there are only two different types of anyons, the vacuum, 1 and the non-Abelian anyon, τ. The only non-trivial fusion rule is

$$\tau \times \tau = 1 + \tau. \qquad (4.47)$$

The quantum dimension of τ can be obtained from $d^2_\tau = d_1 + d_\tau$ giving $d_\tau = \phi$, where $\phi = (1 + \sqrt{5})/2$ is the golden mean. This number has been used extensively by artists, such as the ancient Greek sculptor Phidias or Leonardo Da Vinci in geometrical representations of nature as it describes the ratio that is aesthetically most appealing.

It is interesting to look in detail at all the possible in-between outcomes when fusing n anyons of type τ, as shown in Figure 4.14(a). There we initially fuse the first two anyons, then their outcome is fused with the third τ anyon and so on. To each step i we assign an index e_i that indicates the outcome of the fusion at that step being either 1 or τ. The states $|e_1, e_2, \ldots, e_{n-3}\rangle$ belong to the fusion Hilbert space of the anyons, $\mathcal{M}_{(n)}$. In principle there are 2^{n-3} possible combinations of e_i's, but not all of them are allowed fusion outcomes. Let us analyse how many states $\mathcal{M}_{(n)}$ can have by counting the distinct ways in which one can fuse $n - 1$ anyons of type τ to finally yield a τ. For $n = 1$ we deal with the impossible process where the vacuum turns into a τ anyon, so $\dim(\mathcal{M}_{(1)}) = 0$.

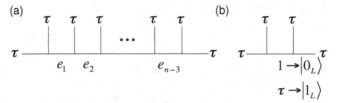

Fig. 4.14   The fusion process for Fibonacci anyons, τ. (a) A series of τ anyons are fused together ordered from left to right. The first two τ anyons are fused and then their outcome is fused with the next τ anyon and so on. (b) Four Fibonacci anyons in state τ created from the vacuum can be used to encode a single logical qubit.

The $n = 2$ case corresponds to a τ as an input and an output going through a trivial process. So $\dim(\mathcal{M}_{(2)}) = 1$. At the next fusing step, the possible outcomes are 1 or τ, giving $\dim(\mathcal{M}_{(3)}) = 1$. When we fuse the outcome with the next anyon then $1 \times \tau = \tau$ and $\tau \times \tau = 1 + \tau$, resulting in two possible τ's coming from two different processes and a single vacuum outcome. So $\dim(\mathcal{M}_{(4)}) = 2$. This signifies that four Fibonacci anyons are needed to encode a qubit. Taking all possible outcomes and fusing them with the next anyon gives a space which is three-dimensional, $\dim(\mathcal{M}_{(5)}) = 3$. Continuing this process one soon notices that the dimension of the fusion space $\dim(\mathcal{M}_{(n)})$ when n anyons of type τ are fused actually reproduces the Fibonacci series,

$$0, 1, 1, 2, 3, 5, 8, 13, \ldots \tag{4.48}$$

It is known from number theory that this dimension is approximately given by the following formula:

$$\dim(\mathcal{M}_{(n)}) \propto \phi^n,$$

in agreement with relation (4.16).

The Fibonacci anyon model can indeed realise universal quantum computation. Much like the Ising model case, the encoding of a qubit can be visualised by employing four τ anyons, as in Figure 4.14(b). There are two distinguishable ways the anyons can be fused that encode the qubit states $|0_L\rangle = |\tau, \tau \to 1\rangle$ and $|1_L\rangle = |\tau, \tau \to \tau\rangle$. To determine the possible quantum gates one needs to evaluate the F and the R matrices. From the fusion rules of Fibonacci anyons and the pentagon identity one finds the non-zero values $(F^{\tau}_{\tau\tau 1})^{\tau}_{\tau} = (F^{\tau}_{1\tau\tau})^{\tau}_{\tau} = (F^{\tau}_{\tau\tau\tau})^{\tau}_{\tau} = (F^{\tau}_{\tau 1\tau})^{\tau}_{\tau} = (F^{1}_{111})^{1}_{1} = 1$ and

$$F^{\tau}_{\tau\tau\tau} = \begin{pmatrix} \frac{1}{\phi} & \frac{1}{\sqrt{\phi}} \\ \frac{1}{\sqrt{\phi}} & -\frac{1}{\phi} \end{pmatrix}. \tag{4.49}$$

These solutions are unique up to a choice of gauge. Inserting these values into the hexagon identity, one obtains the following R matrix describing the exchange of two anyons:

$$R_{\tau\tau} = \begin{pmatrix} e^{4\pi i/5} & 0 \\ 0 & -e^{2\pi i/5} \end{pmatrix}. \tag{4.50}$$

It can be shown that the braiding unitaries $b_1 = R_{\tau\tau}$ and $b_2 = (F^{\tau}_{\tau\tau\tau})^{-1} R_{\tau\tau} F^{\tau}_{\tau\tau\tau}$ acting in the logical space $|0_L\rangle$ and $|1_L\rangle$ are dense in SU(2) in the sense that they can reproduce any element of SU(2) with accuracy ϵ in a number of operations that scales like $O(\text{poly}(\log(1/\epsilon)))$ (Preskill, 2004). For example, an arbitrary one-qubit gate can be performed as follows. Begin from the vacuum and create four anyons labelled τ_1, τ_2, τ_3 and τ_4. Braiding the first and second anyons implements b_1 and braiding the second and third anyons implements b_2. A measurement of the outcome upon fusing τ_1 and τ_2 projects onto $|0_L\rangle$ or $|1_L\rangle$. Similarly, by performing braiding of eight anyons and keeping in mind that $\dim(\mathcal{M}_{(8)}) = 13$ one obtains a dense subset of SU(13). Since SU(4)$\subset$SU(13), we can implement any two-qubit gate (e.g., the CNOT gate) with arbitrary accuracy. This means, the Fibonacci anyon model allows for universal computation on n logical qubits using $4n$ physical anyons (Freedman *et al.*, 2002a).

Summary

In this chapter we introduced the anyonic models in a systematic way and we derived consistency equations between their properties. For example, the spin-statistics theorem relates the spin of anyons with their braiding behaviour. Moreover, the pentagon and hexagon identities can be constructed from simple considerations of anyonic worldline diagrams. These identities establish the fusion and braiding properties of non-Abelian anyons by determining their F and R matrices.

To employ anyons for quantum computation we first identify which part of the fusion Hilbert space is ideal for encoding information. The F and R matrices are then identified as logical gate primitives that non-trivially evolved the fusion states. If the F and R unitary matrices can efficiently span the whole encoding space then the corresponding anyonic model can perform universal quantum computation.

As concrete examples we investigated the Ising and the Fibonacci models. The interest in the Ising anyons is due to their possible physical realisation with near future technology. Nevertheless, this model cannot, per se, support universal quantum computation. Supplementing it with simple dynamical phase rotations can overcome this caveat. On the other hand, the Fibonacci model is universal. Successive applications of its F and R matrices can rotate any state encoded in the fusion space to any other with well-controlled accuracy.

Exercises

4.1 For $\bar{a}$ denoting the antiparticle of the a anyon, demonstrate the following properties of N_{ab}^c: $N_{a1}^c = \delta_{ac}$, $N_{ab}^1 = \delta_{b\bar{a}}$, $N_{ab}^c = N_{ba}^c = N_{b\bar{c}}^{\bar{a}} = N_{\bar{a}b}^{\bar{c}}$ and $\sum_e N_{ab}^e N_{ec}^d = \sum_f N_{af}^d N_{bc}^f$.

4.2   Show that starting from the definition of the quantum dimension (4.15) one can derive the asymptotic relation (4.16). [*Hint*: Consider the matrix N^c with non-negative elements $(N^c)_{ab} = N^c_{ab}$ and decompose it into eigenstates and eigenvalues (Preskill, 2004; Verlinde, 1988).]

4.3   Starting from pairs of Ising anyons created from the vacuum can we generate an entangled state? [*Hint*: See Brennen *et al.* (2009).]

4.4   Show that the F and R matrices of the Fibonacci model given in (4.49) and (4.50) satisfy the pentagon and hexagon identities.

PART II

TOPOLOGICAL MODELS

Quantum double models

The birth of topological quantum computation took place when Alexei Kitaev (2003) made the ingenious step of turning a quantum error correcting code into a many-body interacting system. In particular, he defined a Hamiltonian whose eigenstates are also states of a quantum error correcting code. Beyond the inherited error correcting characteristics, topological systems protect the encoded information with the presence of the Hamiltonian that energetically penalises transformations between states. This opens the door for employing a large variety of many-body effects to combat errors.

Storing or manipulating information with a real physical system is naturally subject to errors. To obtain a reliable outcome from a computation we need to be certain that the processed information remains resilient to errors at all times. To overcome errors we need to detect and correct them. The error detection process is based on an active monitoring of the system and the possibility of identifying errors without destroying the encoded information. Error correction employs the error detection outcome and performs the appropriate steps to correct it, thus reconstructing the original information.

Classical error correction uses redundancy to spread information in many copies so that errors can be detected, for example by majority voting, and then corrected. Similarly, quantum error correction aims to detect and correct errors of stored quantum information. Quantum states cannot be cloned (Wootters and Zurek, 1982), so the repetition encoding cannot be employed. The principle of quantum error correction is to encode information in a sophisticated way that gives the ability to monitor and correct errors. More concretely, the encoding is performed non-locally such that errors, assumed to act in a local way, can be identified and then corrected without accessing the non-local information.

Examples of topological systems that correspond directly to quantum error correction methods are the quantum double models. The Hamiltonian of these models has the encoded states as ground states, such that errors appear as excitations. Ensuring that there is an energy gap above the ground states any error, corresponding to an excited state, will be energetically penalised. Beyond the energetic protection of information, anyonic statistics is intriguingly linked to the error correcting encoding. The non-local characteristic in the encoding of quantum error correction is responsible for the exotic statistics of anyons that emerge as quasiparticle excitations in the quantum double models. As these models enjoy analytic tractability they are a favourable medium to study topological quantum computation.

5.1 Error correction

We start by introducing the main characteristics of quantum error correction. These characteristics provide a useful perspective into the properties of topological systems, even beyond the quantum double models.

5.1.1 Quantum error correcting codes

Quantum error correction is the algorithmic means we have to combat environmental and control errors while performing quantum computation. Similarly to classical error correction, it works on the principle of encoding information in a redundant way. The goal is to perform complex encoding and decoding of information so that the effect of the environment can be neutralised. As we see below, this is possible for particular types of errors.

Let us start with some definitions. Consider a Hilbert space $\mathcal{H}$ of a quantum system spanned by n finite-dimensional complex subsystems $\mathcal{V}$,

$$\mathcal{H} = \mathcal{V} \otimes \dots \otimes \mathcal{V}. \tag{5.1}$$

For simplicity we initially consider $\mathcal{H}$ to be a set of qubits, i.e., $\dim(\mathcal{V}) = 2$. The code space $\mathcal{C}$ is a linear subspace of $\mathcal{H}$ spanned by a subset of its basis states. We encode all the logical information in the code $\mathcal{C}$, while the rest of the Hilbert space is employed to shield this information. Moreover, we define a general k-local operator $\mathcal{O}$ as an operator that acts non-trivially on at most k neighbouring subsystems of $\mathcal{H}$. In this case $\mathcal{O}$ is also known as an operator of length k. Let $\Pi_\mathcal{C}$ be the projector on $\mathcal{C}$. That is, it acts on states in $\mathcal{H}$ and returns their component in $\mathcal{C}$. Consider now the action of the operator $\Pi_\mathcal{C}\mathcal{O}$ on states in $\mathcal{C}$ for any $\lfloor k/2 \rfloor$-local operator, $\mathcal{O}$, where $\lfloor x \rfloor$ denotes the smallest integer that is larger than or equal to x. When applied to a state in $\mathcal{H}$, it returns a state in $\mathcal{C}$ that might not be normalised. In this case $\mathcal{C}$ is called a k-code if

$$\Pi_\mathcal{C}\mathcal{O}\Pi_\mathcal{C} \propto \Pi_\mathcal{C}. \tag{5.2}$$

Condition (5.2) implies that any state in $\mathcal{C}$ can be retrieved after a $\lfloor k/2 \rfloor$-local operator acts on it simply by projecting it again on $\mathcal{C}$. The proportionality symbol $\propto$ indicates that the correct state is retrieved up to an overall normalisation. It has been shown that such a code can effectively protect against errors that act on less than $\lfloor k/2 \rfloor$ qubits (Gottesman, 1998). Such a code is also called $[[n, d, k]]$, where n is the total number of qubits and 2^d is the dimension of the k-code $\mathcal{C}$. This code requires n physical qubits to encode d logical ones protected against errors that are at most $\lfloor k/2 \rfloor$-local.

As an example, consider a set of orthogonal states $|i\rangle$ for $i = 1, \dots, 8$ and operators $\mathcal{O}_r$ that act on them as

$$\mathcal{O}_r |i\rangle = |(i + r) \bmod 8\rangle. \tag{5.3}$$

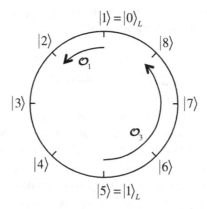

Fig. 5.1 States $|i\rangle$ with $i = 1, \ldots, 8$ arranged on a circle. The operators $\mathcal{O}_r$ for integer r transform these states in a way that respects the periodicity of the circle. A logical qubit is encoded by $|0\rangle_L = |1\rangle$ and $|1\rangle_L = |5\rangle$. Error correction corresponds to bringing the state of the system to the closest logical state. Hence, operator $\mathcal{O}_1$ acting on logical state $|0\rangle_L$ or $|1\rangle_L$ corresponds to a correctable error, unlike errors of the form $\mathcal{O}_2, \mathcal{O}_3, \mathcal{O}_4, \mathcal{O}_5$ and $\mathcal{O}_6$.

In other words we assume that $\mathcal{O}_r$ acts in a periodic way on the states, as shown in Figure 5.1. We can encode the logical qubit states as $|0\rangle_L = |1\rangle$ and $|1\rangle_L = |5\rangle$. Assume that a single error of the form $\mathcal{O}_r$ acts on a qubit state. Our error detecting strategy is to measure the state of the system. If it is different from the logical one then we error correct by taking it to the nearest logical state. Then errors that act as operators $\mathcal{O}_1$ can be efficiently corrected. Errors of the form $\mathcal{O}_2$ cannot be efficiently corrected as we cannot uniquely deduce the initial logical state. Finally, errors $\mathcal{O}_3$, $\mathcal{O}_4$ and $\mathcal{O}_5$ will eventually cause a logical error as the error correction procedure will change the logical state of the qubit. So the code is $[[3, 1, 2]]$ as it requires 2^3 states to encode a single qubit and it protects against 1-local errors. The error detection and correction steps can be performed with conditional gates that do not directly access the encoded information.

5.1.2 Stabiliser codes

Quantum error correcting codes are commonly expressed in the stabiliser formalism (Gottesman, 1997). A stabiliser $\mathbf{T}_n$ is a set of Hermitian operators, T_i with $i = 1, \ldots, n$ that commute with each other, i.e.,

$$[T_i, T_j] = 0 \text{ for all } i, j. \tag{5.4}$$

A particular example of stabilisers can be constructed from the Pauli group, $\mathbf{P}_n$, generated under multiplication by the Pauli matrices $\sigma^x, \sigma^y, \sigma^z$ and the identity $\mathbb{1}$ acting on n qubits. As different Pauli operators acting on the same qubit anticommute, only a subset of the Pauli group operators commute with each other. This subset can form a stabiliser set. The Pauli operators are Hermitian and they square to the identity, so they have eigenvalues ± 1. The Pauli group, $\mathbf{P}_n$, admits a common set of 2^n eigenstates uniquely identified by the pattern of eigenvalues for all stabiliser operators. The stabilised space consists of all eigenstates $| \Psi \rangle$ with particular eigenvalue $+1$ for all operators T_i, i.e.,

$$T_i | \Psi \rangle = | \Psi \rangle \quad \text{for all } i. \tag{5.5}$$

One can define an error correcting code with the stabiliser formalism based on the Pauli group $\mathbf{P}_n$. For that consider a maximal commuting set of independent operators $\mathbf{S}_n$ that is a subgroup of $\mathbf{P}_n$. The group $\mathbf{S}_n$ has n elements that can all be simultaneously diagonalised and have 2^n independent eigenstates. Now let S be a subgroup of $\mathbf{S}_n$. Then we can define a stabiliser code $\mathcal{C}$ to be the set of states that are eigenstates of all elements of S with eigenvalue $+1$ as it applies to the states $| \Psi \rangle$ in (5.5). The basis states of the code space can be parameterised by their eigenvalues with respect to the rest of the elements in $\mathbf{S}_n$ that do not belong to S. Logical information is stored in the code space $\mathcal{C}$. Hence, if S has s elements, it can encode $d = n - s$ qubits, so

$$\dim(\mathcal{C}) = 2^{n-s}. \tag{5.6}$$

Storing quantum information in the code space is useful for probing occurred errors. The stabiliser formalism describes errors with commuting operators so they can be measured and dealt with independently. Suppose that an error operator, which does not commute with S, acts on a code state. This will change the eigenvalue of the state for some of the operators in S. As any state in $\mathcal{C}$ is an eigenstate of all operators in S with eigenvalue $+1$, we can measure these operators without changing the computational state (i.e., without reading out information). In this way we can measure whether an error occurred or not and correct for it without interrupting the computation.

Let us now consider the centraliser $Z(S)$, which is the maximal set of operators that commutes with S. Elements of $Z(S)$ do not move code states out of $\mathcal{C}$. In other words, their action on $\mathcal{C}$ gives states that are also eigenstates of the S operators with eigenvalues $+1$. Hence, they perform transformations between code states. Such elements serve as encoded logical operations. If, on the other hand, an error belongs to $Z(S)$, then it will be undetected and will destroy the encoded information. Table 5.1 summarises the main characteristics of quantum error correction.

When we try to error correct a detected error then the composite operator of the error and the error correction operation need to be an identity in order to bring back the state to its original form. If this composite operator is a non-trivial element of $Z(S)$, it causes an irreversible error. Which of the two cases takes place depends on the length of the error in relation to the distance of the code. The distance k of the code $\mathcal{C}$ is therefore the minimal

Table 5.1	Quantum error correction
$\mathbf{P}_n$	Pauli group acting on n qubits
$\mathbf{S}_n$	Maximal set of commuting $\mathbf{P}_n$ elements
S	A subgroup of $\mathbf{S}_n$
$\mathcal{C}$	Set of states $| \psi \rangle$ with $T_i | \psi \rangle = | \psi \rangle$ for all $T_i \in S$ (*encoding states*)
$Z(S)$	Maximal set of $\mathbf{P}_n$ elements that commute with S (*logical operations*)

length among the elements of $Z(S)$ that do not belong in S. For an efficient encoding we assume that the errors are less than $\lfloor k/2 \rfloor$-local so that error correction procedures can be found that return the state back to its original form (Gottesman, 1998). Interestingly, these concepts take a transparent geometrical interpretation when they are applied to topological systems.

5.2 Quantum double models

Quantum double models are particular lattice realisations of topological systems. They are based on a finite group, G, that acts on spin states, defined on the links of the lattice. Based on these groups, stabiliser Hamiltonians can be defined consistently that have analytically tractable spectra. It can be shown that the ground states of these Hamiltonians behave like error correcting codes. Anyons are associated with properties of the spin states around each vertex or plaquette of the lattice. The fusion and braiding behaviour of the anyons depends on the property of the employed group, G. For example, an Abelian group leads to Abelian anyons and a non-Abelian group leads to non-Abelian anyons. All the properties of the anyons emerge from the mathematical structure of the quantum double, denoted $D(G)$. In the following we present a simple example of quantum doubles, the Abelian toric code.

5.2.1 The toric code

The simplest quantum double model is the toric code (Kitaev, 2003), denoted by $D(Z_2)$. It is based on the group $Z_2 = \{0, 1\}$, with $0 \cdot 0 = 0$, $1 \cdot 1 = 0$ and $0 \cdot 1 = 1$, that acts on spin-1/2 particles, which are defined on the links of a lattice. The Hamiltonian of this system is defined in terms of Pauli operators. By employing simple properties of Pauli operators we show that the toric code supports Abelian anyons. Due to its simplicity the toric code is one of the most studied topological models. It can serve as a platform for various quantum information tasks and as a test bed for the properties of topological systems.

5.2.1.1 Hamiltonian

Consider a square lattice with qubits or spin-1/2 particles positioned at the lattice links, as shown in Figure 5.2. To construct the Hamiltonian we employ the vertex and plaquette interaction terms given, respectively, by

$$A(v) = \sigma^x_{v,1}\sigma^x_{v,2}\sigma^x_{v,3}\sigma^x_{v,4} \tag{5.7}$$

and

$$B(p) = \sigma^z_{p,1}\sigma^z_{p,2}\sigma^z_{p,3}\sigma^z_{p,4}. \tag{5.8}$$

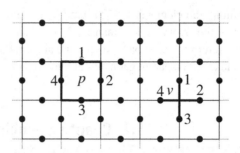

Fig. 5.2   The toric code defined on a square lattice with spin-1/2 particles positioned at its edges. The interaction terms of the model are the vertex operators $A(v) = \sigma^x_{v,1}\sigma^x_{v,2}\sigma^x_{v,3}\sigma^x_{v,4}$, where the enumeration runs around the spins of vertex v and the plaquette operators $B(p) = \sigma^z_{p,1}\sigma^z_{p,2}\sigma^z_{p,3}\sigma^z_{p,4}$, where the enumeration runs around the four spins of plaquette p.

The indices $1, \ldots, 4$ of the Pauli operators, σ^z or σ^x, enumerate the edges of each vertex or plaquette. The defining Hamiltonian is

$$H = -\sum_v A(v) - \sum_p B(p). \tag{5.9}$$

Each of the interaction terms commutes with the Hamiltonian as well as with each other. Moreover, they square to one so their eigenvalues are ± 1. Using these properties, we can find the spectrum of Hamiltonian (5.9). In particular, its ground state is given by

$$|\xi\rangle = \prod_v \frac{1}{\sqrt{2}}(\mathbb{1} + A(v))|00\ldots 0\rangle \tag{5.10}$$

since $\sigma^z|0\rangle = |0\rangle$ for all spins of the lattice. Indeed, $|\xi\rangle$ is an eigenstate of all $A(v)$ and $B(p)$ operators with eigenvalue 1.

5.2.1.2 Anyons and their fusion

The ground state $|\xi\rangle$ corresponds, by definition, to the anyonic vacuum, i.e., the absence of any anyon. One can excite pairs of anyons on the lattice using single spin operations. By applying σ^z rotations on a spin of the lattice a pair of quasiparticle excitations is created on the two neighbouring vertices, as shown in Figure 5.3. These quasiparticles correspond to eigenvalue -1 for the corresponding $A(v)$ operators and are called e-type anyons. The state of these anyons is usually denoted as $|e, e\rangle = \sigma^z|\xi\rangle$. It describes an e anyon positioned at each vertex neighbouring the rotated spin. An m-type anyon lives on plaquettes for which the $B(p)$ operator has eigenvalue -1. The m anyons are created in pairs from σ^x rotations such that $|m, m\rangle = \sigma^x|\xi\rangle$. The combination of both e and m anyons creates the composite quasiparticle ϵ with $|\epsilon, \epsilon\rangle = \sigma^z\sigma^x|\xi\rangle = i\sigma^y|\xi\rangle$.

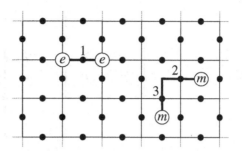

Fig. 5.3 By applying a σ^z rotation on spin 1 of the ground state $|\,\xi\,\rangle$, the operators $A(v)$ of the vertices neighbouring spin 1 obtain eigenvalue -1. This detects the presence of e anyons. Two σ^x rotations create four m anyons. If two m anyons are positioned at the same plaquette (e.g., due to the σ^x rotations of spins 2 and 3) they annihilate each other. This finally gives two anyons at the endpoints of the string passing though the rotated spins 2 and 3.

The presence of e, m and ϵ quasiparticle excitations is detected by measuring the eigenvalues of the corresponding $A(v)$ or $B(p)$ operators. Eigenvalue $+1$ corresponds to the vacuum, while -1 detects the presence of an anyon. If the same Pauli rotations are applied on spins of the same vertex or plaquette then they create two overlapping anyons. The resulting eigenvalue of $A(v)$ or $B(p)$, respectively, is $+1$. So the outcome of the fusion is the vacuum, as shown in Figure 5.3. By time-ordering the Pauli rotations we can move anyons around the lattice. We shall indicate the position of the rotations with a string. The anyons are always at its endpoints. Strings associated with e anyons lie on the square lattice, while strings of m anyons lie on its dual square lattice. The dual lattice has its vertices at the centre of the plaquettes of the original lattice. For a system with open boundaries a string may end up at the boundary, describing a single anyon at its free endpoint.

In general, an even number of σ^z rotations applied to the spins of a certain vertex, v, has eigenvalue of $A(v)$ equal to 1, while an odd number has eigenvalue -1. Similarly, an even number of σ^x rotations on spins of the same plaquette gives eigenvalue 1, while an odd number gives -1. Together with the composition rule for the ϵ particle, these properties translate to the following fusion rules:

$$e \times e = m \times m = \epsilon \times \epsilon = 1, \; e \times m = \epsilon, \; \epsilon \times e = m \text{ and } \epsilon \times m = e \qquad (5.11)$$

that describe the outcome from combining two anyons, where 1 is the vacuum state. The fusion rules dictate that if two anyons are created on the same plaquette or vertex, then they annihilate. The annihilation operation also glues two single strings of spin rotations of the same type together to form a longer string, again with a pair of anyons at its ends, as shown in Figure 5.3.

If a string of σ^x or σ^z operations forms a loop, then the anyons at its ends annihilate each other, thus removing anyonic excitation. An isolated elementary loop of σ^z rotations around a plaquette p is actually the $B(p)$ operator. If we span a loop around two neighbouring plaquettes p and p', as shown in Figure 5.4, then the resulting operator corresponds to $B(p)B(p')$. In other words, a larger loop can be broken down into smaller ones that overlap at the internal edges, as $(\sigma^z)^2 = 1$. Similarly, loops of σ^x operators can be constructed out of $A(v)$ operators. The eigenvalue of the σ^x loop operators detects if the loop encloses an

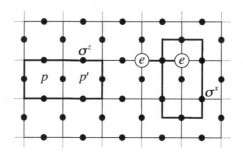

Fig. 5.4   A loop of σ^z operators along the lattice can be built from the product of neighbouring $B(p)$ operators on plaquettes p and p'. A loop of σ^x operators along the edges of the dual lattice also results from the product of $A(v)$ operators. These loop operators can detect the parity of the total e or m anyons enclosed by the loops.

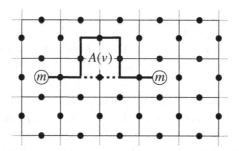

Fig. 5.5   A straight string of three σ^x rotations creates two m anyons at its endpoints. When an $A(v)$ operation is applied at a vertex neighbouring the string then their common site cancels and the string of σ^x operators is deformed.

odd or an even number of e anyons. Differently put, they read the total fusion outcome of all the enclosed e anyons. Equivalent considerations hold for σ^z loops and the corresponding m anyons.

Finally, we would like to study the properties of the strings that connect anyons. Consider two m anyons at the endpoints of a string of σ^x operators. Their state is invariant with respect to deformations of the shape of the string as long as its endpoints remain fixed. This can be verified in the following way. The state of the two anyons does not change if we apply any of the stabilisers $A(v)$. If the elementary square loop that corresponds to $A(v)$ has a common edge with the string operator, then this edge-operator will be cancelled and the shape of the string will be deformed, as shown in Figure 5.5. Hence, both strings, the straight and the deformed one, give rise to the same two-anyon state.

An alternative approach is to consider the ground state $|\xi\rangle$. From (5.3) we see that the ground state is the equal superposition of all possible products of elementary loops $A(v)$. Application of any contractible loop operator on the ground state gives back the same state with its components rearranged. The excited states inherit this property by being the equal superposition of all possible strings that connect the two anyons. Hence, the exact shape of the string does not have any physical meaning. Only the position of the anyons does.

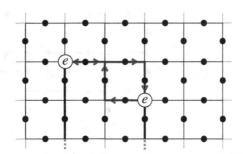

Fig. 5.6 Exchanging two e anyons by transporting them along different paths. The final configuration has an additional loop operator that has trivial action as it acts on empty plaquettes. Hence, the e anyons have bosonic mutual statistics.

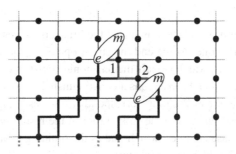

Fig. 5.7 The exchange of two ϵ particles by subsequent exchanges of their constituent e and m anyons. Care needs to be taken not to rotate the ϵ particles between the initial and final states to avoid contributions from spin rotations. The exchange operation is given by a sequence of Pauli rotations on qubits 1 and 2, i.e., $(\sigma_1^x \sigma_1^z \sigma_2^z \sigma_2^x)(\sigma_2^x \sigma_2^z \sigma_1^x \sigma_1^z) = -1$. Hence, the ϵ's are fermions.

5.2.1.3 Anyonic statistics

We now probe the statistical behaviour of e, m and ϵ quasiparticles. Consider two e anyons, as shown in Figure 5.6. We can exchange their position by applying σ^z rotations. The final configuration is equal to the initial one with the addition of a loop of σ^z's passing through the position of both anyons. As this loop operator acts on plaquettes with no m anyons, it gives back the identity. Hence, the final state of the system equals the initial one, thereby signalling the bosonic mutual statistics of e anyons.

The same argument holds for the m anyons, but not for the ϵ anyons. We can actually demonstrate that the mutual statistics of ϵ particles is fermionic. Consider a pair of ϵ's constructed out of the constituent e and m anyons, as presented in Figure 5.7. The anyon ϵ is an extended object as it occupies a plaquette and its neighbouring vertex. We would like to exchange the positions of the ϵ's without rotating them. An overall rotation could cause extra phase factors due to the spin particle ϵ might have. Under this condition an exchange is given by the following Pauli rotations acting on the state of the two fermions:

$$(\sigma_1^x \sigma_1^z \sigma_2^z \sigma_2^x)(\sigma_2^x \sigma_2^z \sigma_1^x \sigma_1^z) = -1. \tag{5.12}$$

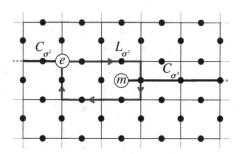

Fig. 5.8 String operators C_{σ^z} and C_{σ^x} have e and m anyons at their endpoints, respectively. A looping operation L_{σ^z} braids the e anyon around the m and brings it back to its original position. This braiding gives rise to a non-trivial phase factor due to the anticommutation of σ^x and σ^z operators of the strings that meet at a single point.

Each set of spin rotations inside the brackets moves each particle from its initial position to the final one. At the end the string operators cancel each other and an overall phase of -1 remains, which reveals the fermionic statistics of ϵ particles.

As the e and the m anyons are distinguishable, we cannot directly exchange them, but we can braid them. Braiding corresponds to two exchanges from where we can attribute their exchange statistics as the square root of the resulting evolution. We now show that a non-trivial phase emerges, which makes the braiding evolution similar to the Aharonov–Bohm effect. Consider the initial state

$$|\Psi_{\text{ini}}\rangle = C_{\sigma^x} C_{\sigma^z} |\xi\rangle, \tag{5.13}$$

where C_{σ^α}, with $\alpha = x$ or z, are strings of Pauli operators with the corresponding anyons at their endpoints. This gives rise to neighbouring e and m anyons, as shown in Figure 5.8. During braiding, the e anyon is moved around the m anyon along the path generated by successive applications of σ^z rotations, thereby giving rise to the looping L_{σ^z} operator. Note that the L_{σ^z} operator meets the C_{σ^x} across a single spin, so

$$L_{\sigma^z} C_{\sigma^x} = -C_{\sigma^x} L_{\sigma^z}. \tag{5.14}$$

We can now manipulate the final state in order to bring the operator L_{σ^z} to act on the vacuum, where we know it has trivial action. Then the final state is given by

$$|\Psi_{\text{fin}}\rangle = L_{\sigma^z} |\Psi_{\text{ini}}\rangle = L_{\sigma^z}(C_{\sigma^x} C_{\sigma^z} |\xi\rangle) = -C_{\sigma^x} C_{\sigma^z} L_{\sigma^z} |\xi\rangle = -|\Psi_{\text{ini}}\rangle. \tag{5.15}$$

This result is independent of the particular shape of the loop L_{σ^z} as long as it circulates the m anyon exactly once. The topological phase factor of -1 reveals a non-trivial statistics between e and m anyons. Symbolically, we assign the exchange phase i to these anyons. This behaviour is different from the braiding of bosons or fermions. The same phase is obtained if we braid e or m with ϵ particles. It is worth noting that the -1 in (5.12) corresponds to a single exchange and signals the fermionic character of ϵ, while the -1 in (5.15) corresponds to two exchanges and signals the anyonic character between different types of particles of the toric code.

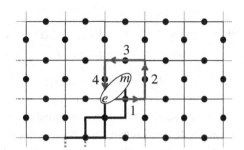

Fig. 5.9 Rotating counterclockwise an ϵ fermion by 2π around itself is equivalent to moving its constituent e anyon around the m one by σ^z rotations on the qubits 1, 2, 3 and 4. The resulting phase factor -1 reveals that ϵ has spin one-half.

We would now like to probe the spin of these particles. The e and m particles are bosons and have trivial spins. In particular, they cannot be rotated as they do not have a sense of orientation. The ϵ particles are expected to have spin one-half as they are fermions. Indeed, a counterclockwise 2π rotation corresponds to moving the e constituent anyon around m by the operator $\sigma_1^z \sigma_2^z \sigma_3^z \sigma_4^z$. This is a plaquette operator that detects the population of m's leading to the phase factor of -1, thereby revealing the half-spin of ϵ fermions (Rauch *et al.*, 1975), as shown in Figure 5.9.

5.2.1.4 Encoding information

Previously we have demonstrated that the quasiparticle excitations of the toric code have anyonic statistics. These properties do not depend on the topology of the surface where the lattice Hamiltonian is defined. The surface topology becomes, however, important when one wants to employ the toric code to encode quantum information. A non-trivial genus can give rise to ground state degeneracy. For concreteness consider a torus with genus one. A torus configuration can be produced by imposing periodic boundary conditions in both directions of a toric code lattice, as given in Figure 5.2. Transforming one ground state to another involves creating a pair of anyons and then moving them along non-contractible loops of the torus before re-annihilating them, as shown in Figure 5.10. As the resulting state does not have excitations, it corresponds to a ground state. Moreover, it is different from the initial state as the loop is non-contractible. Denoting the two non-equivalent trajectories on the torus as 1 and 2, we can define the ground states

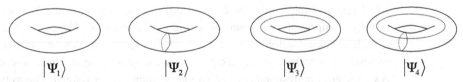

$$|\Psi_1\rangle \qquad\qquad |\Psi_2\rangle \qquad\qquad |\Psi_3\rangle \qquad\qquad |\Psi_4\rangle$$

Fig. 5.10 The torus with genus $g = 1$ with the toric code lattice Hamiltonian defined on it (not depicted). The ground state has fourfold degeneracy $|\Psi_i\rangle$, for $i = 1, \ldots, 4$. Starting from the vacuum state $|\Psi_1\rangle$ one can create a pair of e anyons and wrap them around two non-trivial inequivalent loops on the torus giving the states $|\Psi_2\rangle$ and $|\Psi_3\rangle$. State $|\Psi_4\rangle$ corresponds to spanning both inequivalent loops.

$$|\Psi_1\rangle, \; |\Psi_2\rangle = C_{\sigma^z}^1 |\Psi_1\rangle, \; |\Psi_3\rangle = C_{\sigma^z}^2 |\Psi_1\rangle, \; |\Psi_4\rangle = C_{\sigma^z}^1 C_{\sigma^z}^2 |\Psi_1\rangle. \tag{5.16}$$

The state $|\Psi_1\rangle$ can be chosen to be the ground state $|\xi\rangle$ given explicitly in (5.10). Then the operators $C_{\sigma^z}^1$ and $C_{\sigma^z}^2$ act on $|\xi\rangle$ by generating a pair of e anyons, moving them along the directions 1 or 2 respectively and then annihilating them. The resulting states are invariant under continuous deformations of the anyonic loops. So only four states can be created in this way. These states are linearly independent as they correspond to superpositions of loop configurations that differ with respect to their winding around the torus. A linearly dependent set of states can be obtained by employing the loop operators that correspond to m or combinations of e and m anyons (see Exercise 5.3). Hence, a four-dimensional Hilbert space arises that can encode two qubits. If the toric code is defined on a surface with genus g, then it can encode $2g$ qubits.

The fourfold degeneracy of the toric code ground state can be verified independently by considering the number of qubits in the system and by counting the number of stabiliser conditions imposed on the ground state. On a square lattice, $L \times L$, we have $2L^2$ spins. The stabiliser conditions are

$$A(v)|\xi\rangle = |\xi\rangle \;\; \text{and} \;\; B(p)|\xi\rangle = |\xi\rangle \tag{5.17}$$

for all L^2 vertices v and for all L^2 plaquettes p. With open boundaries the stabiliser conditions are as many as the spins, so they define a single ground state, which is given in (5.10). If we impose periodic boundary conditions on the lattice (i.e., restrict it on a torus) then the stabiliser conditions are not all independent. We can verify that the following conditions hold:

$$\prod_v A(v) = \mathbb{1} \;\; \text{and} \;\; \prod_p B(p) = \mathbb{1}. \tag{5.18}$$

In other words, multiplying all $A(v)$ or all $B(p)$ operators together on the torus involves the squaring of Pauli operators as there are always overlapping edges. Hence, there are two stabiliser conditions less, one from the A's and one from the B's. Equation (5.17) therefore has four solutions rather than one, in agreement with the number of code states suggested by (5.6).

Now we are in a position to interpret the toric code as an error correcting code. Consider the toric configuration with L spins at each of the two directions where we impose periodic boundary conditions. The L^2 spins comprise the Hilbert space $\mathcal{H}$. The operators $A(v)$ and $B(p)$ comprise the stabiliser set of commuting operators. Hence, the degenerate ground state, here being fourfold, is the code $\mathcal{C}$, where logical information can be encoded. Errors correspond to undesired spin rotations that, as we have seen, create anyonic excitations. A k-local error corresponds to k neighbouring qubits being rotated. The worst case scenario is having an extended string operator of length k with two anyons at its endpoints. String operators like $C_{\sigma^z}^1$ create a pair of anyons, move one of them around a non-contractible loop and re-annihilate them. This operation corresponds to encoded logical gates. Error detection corresponds to measuring the errors with the help of the stabilisers that can

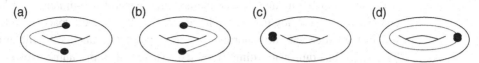

Fig. 5.11 Two errors on the toric code in the form of anyonic excitations could have been originated from two different strings, such as the ones depicted in (a) and (b). To fuse the anyonic errors we can move them rightwards or leftwards. Depending on the original path that created the anyons, the resulting configuration might return the system to its correct encoding state (c) or can result in an undesired looping operator (d).

recognise the position of the anyons. Error correction then corresponds to fusing these anyons together, which returns the system back to the vacuum. Nevertheless, we need to make sure we do not create non-contractible loops during the error correction procedure, since that would correspond to an undesired logical gate.

To analyse the error correction process, consider the toric code with two anyonic errors. In Figure 5.11(a) and (b) we depict two different ways these errors may have been created. The error detection process can measure the position of the anyons. However, the string that generated the anyons is not measurable, so we cannot distinguish between these two cases. The error correction step aims to fuse the two anyons so that the state of the system becomes a ground state. Nevertheless, we would like to recover the original state before the errors happened and not perform a logical error by realising a non-contractible loop. These two cases are described in Figure 5.11(c) and (d).

The strategy we adopt is to annihilate the anyons through the shortest possible path on the geometry of the torus. This elimination of errors could affect the logical space only if the two errors have propagated a distance larger than $\lfloor L/2 \rfloor$, where L is the linear size of the torus. In that case the error correction step might result in a non-contractible loop that corresponds to an undesired logical gate. Otherwise, the shortest distance annihilation method can protect against any error that is $\lfloor L/2 \rfloor$-local. Hence, the toric code corresponds to a $[[L^2, 2, L]]$ error correcting code.

With the toric code Hamiltonian (5.9) present, the generation of errors is penalised by an energy gap. Errors in the form of virtual anyonic excitations are exponentially suppressed from going around the torus. Hence, large torus size is favourable. Such Abelian models are not suited for quantum computation as the encoded logical gates produce only phase factors and, thus, are not universal. But we can employ the Abelian models as quantum memories. The general case of Abelian anyons is analysed in Example I later in this chapter.

5.2.2 General $D(G)$ quantum double models

We now discuss the quantum double models $D(G)$ for a general finite group G. The Hamiltonian of these models can support a rich variety of anyonic excitations with Abelian and non-Abelian statistics (Bais *et al.*, 1992; Kitaev, 2003). Here we give a general treatment, while Examples I and II below provide a more detailed analysis.

Quantum double models can be defined on a general two-dimensional lattice, but for convenience we employ the square lattice. A spin-like Hilbert space $\mathcal{V}$ is defined at each link of the lattice with orthonormal basis $\{|g\rangle : g \in G\}$, labelled by the group elements. This space has dimension $\dim(\mathcal{V}) = |G|$, where $|G|$ is the number of elements in the finite group G. In the case of the toric code, $\mathcal{V}$ is two-dimensional as Z_2 has only two elements.

To proceed we define the linear operators $L^g_\pm$ with $g \in G$, that are associated with vertices of the lattice and $T^h_\pm$ with $h \in G$, that are associated with plaquettes. These operators act on the spin Hilbert space, $\mathcal{V}$, such that

$$L^g_+ |z\rangle = |gz\rangle, \quad L^g_- |z\rangle = \left|zg^{-1}\right\rangle, \quad T^h_+ |z\rangle = \delta_{h,z} |z\rangle, \quad T^h_- |z\rangle = \delta_{h^{-1},z} |z\rangle. \quad (5.19)$$

In the case of the toric code all the $L^g_\pm$'s correspond to the σ^x Pauli operator, while the $T^h_\pm$'s become $(\mathbb{1} \pm \sigma^z)/2$. The action of the $L^g_\pm$ operators on the states follows the group multiplication, while the $T^h_\pm$'s act as projectors. More concretely, one can show that these operators satisfy the following commutation relations:

$$L^g_+ T^h_+ = T^{gh}_+ L^g_+, \quad L^g_- T^h_+ = T^{hg^{-1}}_+ L^g_-, \quad L^g_+ T^h_- = T^{hg^{-1}}_- L^g_+, \quad L^g_- T^h_- = T^{gh}_- L^g_-. \quad (5.20)$$

To consistently define the Hamiltonian of the system we introduce an orientation in the edges of the lattice. Changing the orientation of an edge corresponds to changing the basis state from $|g\rangle$ to $\left|g^{-1}\right\rangle$. For the toric code case with $Z_2 = \{0, 1\}$ a choice of orientation is not necessary, as $0^{-1} = 0$ and $1^{-1} = 1$. For simplicity we take the vertical edges of the square lattice oriented upwards and the horizontal ones rightwards, as shown in Figure 5.12. To each vertex v of the lattice we assign a vertex operator defined by

$$A(v) = \frac{1}{|G|} \sum_{g \in G} L^g_{+,1} L^g_{+,2} L^g_{-,3} L^g_{-,4}, \quad (5.21)$$

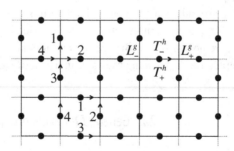

Fig. 5.12   A representation of a quantum double model $D(G)$ defined on a square lattice with $|G|$-dimensional spins at its links. The lattice is oriented with vertical links pointing upwards and horizontal ones pointing rightwards. The signs of the $L^g_\pm$ and $T^h_\pm$ operators are given according to their position with respect to an oriented edge.

with the convention of signs and the enumerations as in Figure 5.12. Similarly, for a plaquette p one can define

$$B(p) = \sum_{h_1 \ldots h_4 = 1} T^{h_1}_{-,1} T^{h_2}_{-,2} T^{h_3}_{+,3} T^{h_4}_{+,4}. \tag{5.22}$$

These operators are local and have a discrete spectrum.

The quantum double model can be viewed in terms of a gauge theory. Let us interpret the transformations $A_g(v) = L^g_{+,1} L^g_{+,2} L^g_{-,3} L^g_{-,4}$ acting on the spin states as local gauge transformations. From this perspective $A(v)$ symmetrises these states with respect to all possible group G transformations. So it can project out states that are not gauge-invariant at vertex v, i.e., it identifies the absence of charges on that vertex. Similarly, the $B(p)$ operator projects out states with non-vanishing magnetic flux passing through plaquette p. Note that the charge and flux description is effectively emerging from the interpretation of the spin states around the vertex or the plaquette. The e and the m anyons of the toric code are therefore also called charge and flux anyons, respectively.

All of the operators $A(v)$ and $B(p)$ commute with each other (see Exercise 5.1). Hence, the Hamiltonian

$$H = -\sum_v A(v) - \sum_p B(p) \tag{5.23}$$

is in the stabiliser formalism and can be diagonalised directly. The ground state $|\xi\rangle$ satisfies

$$A(v)|\xi\rangle = |\xi\rangle, \quad B(p)|\xi\rangle = |\xi\rangle, \text{ for all } v \text{ and } p, \tag{5.24}$$

denoting the absence of anyons. Excitations are identified by the violation of conditions (5.24). They correspond to quasiparticles that live on the vertices or the plaquettes of the lattice, or simultaneously on a vertex and a neighbouring plaquette. It is possible to find the projectors that identify the type of quasiparticles and their properties, but this problem is complex in its generality (Kitaev, 2003). The quasiparticles can be Abelian, if they are based on an Abelian group or non-Abelian, arising for example from the S_3 group that has non-commuting elements. Both cases are presented in detail in the following examples.

The main property of non-Abelian anyonic Hamiltonians that is of interest for quantum computation is their large fusion space degeneracy. This degeneracy is created by the presence of non-Abelian anyons. Hamiltonian (5.23) is naturally gapped as the $A(v)$ and $B(p)$ operators have a discrete spectrum, so quantum information encoded in anyons is energetically protected from errors. The advantage over the Abelian anyon encoding, as we have seen for example with the toric code, is that one can manipulate the information by braiding the anyons together, rather than by creating anyons and circulating them around the torus. In addition, the dimension of the encoding space can be increased by creating more anyons rather than changing the topology in terms of the surface genus. This dramatically simplifies the control procedure and can give rise to universal quantum computation for certain types of non-Abelian models.

5.3 Example I: Abelian quantum double models

The toric code belongs in the general family of cyclic Abelian quantum double models denoted $D(Z_d)$ (Kitaev, 2003). For these models the elements of the group are given by

$$Z_d = \{0, 1, \ldots, d-1\}. \tag{5.25}$$

The product between two group elements $h, g \in Z_d$ is defined as their addition modulo d, i.e.,

$$g \cdot h = g + h \; (\mathrm{mod}\; d). \tag{5.26}$$

Next we consider a lattice with square geometry and assign d-level spins on every edge. We parameterise the spin states by the group elements (5.25). Rotations of the spins are given in terms of the generalised Pauli operators

$$X = \sum_{h \in Z_d} |h+1 \; (\mathrm{mod}\; d)\rangle \langle h|, \; Z = \sum_{h \in Z_d} \omega^h |h\rangle \langle h|, \tag{5.27}$$

where $\omega = e^{i2\pi/d}$. For $d = 2$ we obtain the usual anticommuting Pauli operators, σ^x and σ^z, respectively. In general, they satisfy the commutation relation

$$ZX = \omega XZ. \tag{5.28}$$

As the X operator displaces the labelling of the states by a unit, in a periodic fashion, its eigenstates are given by

$$|\tilde{g}\rangle = \frac{1}{\sqrt{d}} \sum_{h \in Z_d} \omega^{gh} |h\rangle \; \text{ for } g = 0, \ldots, d-1, \tag{5.29}$$

with the corresponding eigenvalues being $\omega^{-g} = e^{-i2\pi g/d}$ for each $g \in Z_d$.

Similarly to the toric code we define the vertex and plaquette operators by

$$A(v) = X_1^\dagger X_2^\dagger X_3 X_4 \text{ and } B(p) = Z_1^\dagger Z_2 Z_3 Z_4^\dagger, \tag{5.30}$$

respectively, where the numbering proceeds as in Figure 5.13. Both these operators have eigenvalues ω^g, for $g = 0, \ldots, d-1$. Note that an orientation of the edges is defined throughout the lattice that assigns the conjugation of the appropriate Pauli matrices in definitions (5.30).

Consider a general eigenstate, $|\psi\rangle$, of all vertex and plaquette operators. We define a certain vertex, v, or plaquette, p, to be unoccupied if $A(v)|\psi\rangle = |\psi\rangle$ or $B(p)|\psi\rangle = |\psi\rangle$, respectively. An anyon e^g is associated with a vertex, v, if $A(v)|\psi\rangle = \omega^g |\psi\rangle$. An anyon m^h is associated with a plaquette, p, if $B(p)|\psi\rangle = \omega^h |\psi\rangle$. The presence of both anyons, e^g and m^h, in adjacent plaquette and vertex is associated with the composite particle $\epsilon^{g,h}$. Following these definitions we can specify the Hamiltonian

$$H = -\left[\sum_{v} \sum_{h \in Z_d} \left(A(v)\right)^h + \sum_{p} \sum_{h \in Z_d} \left(B(p)\right)^h \right] \tag{5.31}$$

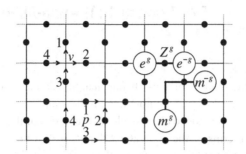

Fig. 5.13   An Abelian quantum double model defined on a square lattice with a d-level spin on each edge. The numbering of the spins around each vertex, v, and plaquette, p, is given together with the link orientations. Anyons e^g reside on vertices and m^g on plaquettes. They are created in pairs and transported by single-spin operations.

that has the anyonic vacuum, $|\xi\rangle$, as its ground state. This Hamiltonian assigns equal energy to all e^g quasiparticle excitations, because the sums $\sum_{h \in Z_d} \left(A(v) \right)^h$ are projectors onto the vacuum state for each vertex and they act identically on each anyon e^g for $g \in Z_d$. An analogous argument applies to plaquette occupations m^g. The resulting anyon model comprises d^2 different particle species given by

$$1, \ e^g, \ m^g, \ \epsilon^{g,h} \text{ for all } g, h \in Z_d, \tag{5.32}$$

identified by the $A(v)$ and $B(p)$ operators. The fusion rules of these particles are given by

$$e^g \times e^h = e^{g+h \, (\mathrm{mod} \, d)}, \ m^g \times m^h = m^{g+h \, (\mathrm{mod} \, d)} \text{ and } e^g \times m^h = \epsilon^{g,h}. \tag{5.33}$$

Consider now the braiding operation of an e^g around an m^h. From the commutation relation (5.28) we deduce the R matrix

$$(R^{\epsilon^{g,h}}_{e^g \, m^h})^2 = \omega^{gh}, \text{ where } \omega = e^{i2\pi/d}. \tag{5.34}$$

The rest of the non-trivial braiding matrices are deduced from $R^{\epsilon^{g,h}}_{e^g \, m^h}$. As these are Abelian anyons, all their F matrices are trivial.

The presence of anyons corresponds to a violation of the stabiliser conditions of the vertex or plaquette operators. The generation of anyons is achieved by applying Z or X spin rotations to the ground state $|\xi\rangle$, as shown in Figure 5.13. Due to the fusion relations (5.33), the antiparticle of e^g is e^{-g} and the antiparticle of m^g is m^{-g}. In other words, single-spin rotations create particle–antiparticle pairs with positions that are determined by the orientation of the corresponding link. A Z^g rotation of spins 1 or 2 of a vertex, v, or Z^{-g} rotation on 3 or 4, creates an e^g charge at that vertex and an e^{-g} charge on the other vertex shared by the rotated spin. Similarly, an X^g on spins 2 or 3 of a plaquette, p, or a X^{-g} on 1 or 4, creates an m^g flux on that plaquette and an m^{-g} on the other plaquette shared by the rotated spin. In general, each string of Z or X rotations has a particle on one end and its antiparticle on the other.

5.4 Example II: The non-Abelian $D(S_3)$ model

Since the structure of non-Abelian quantum double models $D(G)$ is most easily approached through an example, we now look at the simplest non-Abelian finite group $G = S_3$. This group corresponds to all possible permutations of three objects. We express every element of S_3 in terms of the generator t that describes the exchange of two specific objects out of the three, and the generator c that yields a cyclic permutation of all three objects along a given direction. These operators satisfy $t^2 = c^3 = e$ and $tc = c^2t$, where e denotes the identity element. Using this notation we find six independent elements of the group given by

$$S_3 = \{e, c, c^2, t, tc, tc^2\}, \tag{5.35}$$

so $|S_3| = 6$. Let us pick an oriented two-dimensional square lattice. On each edge there resides a six-level spin spanned by the states $|g\rangle$, where g is an element of S_3.

Define a set of six operators acting on vertex v by

$$A_g(v) = L^g_{+,1} L^g_{+,2} L^g_{-,3} L^g_{-,4}, \quad \text{for } g \in S_3, \tag{5.36}$$

as depicted in Figure 5.14, where the action of the operators $L^g_\pm$ on the edge spins is given in (5.19). These operators satisfy $[A_g(v), A_{g'}(v')] = 0$ for all v and v' as well as g and g' (see Exercise 5.1). Similar definitions hold for the plaquette operators but we shall not give them explicitly here. According to (5.21) we can build the vertex operator $A(v)$ as

$$A(v) = \frac{1}{6}\Big[A_e(v) + A_c(v) + A_{c^2}(v) + A_t(v) + A_{tc}(v) + A_{tc^2}(v)\Big]. \tag{5.37}$$

This operator identifies if a state $|\psi\rangle$ of the whole lattice has the vacuum at vertex v through the condition $A(v)|\psi\rangle = |\psi\rangle$. Hence, we define the vertex operator

$$P_1(v) = A(v) \tag{5.38}$$

as the projector onto the vacuum 1 for vertex v. The stabiliser space consists of states with no quasiparticles, i.e., those for which $A(v)|\xi\rangle = |\xi\rangle$ for all v and $B(p)|\xi\rangle = |\xi\rangle$ for all p, where $B(p)$ is defined as in (5.22). The Hamiltonian of the system is given by

$$H = -\sum_v A(v) - \sum_p B(p). \tag{5.39}$$

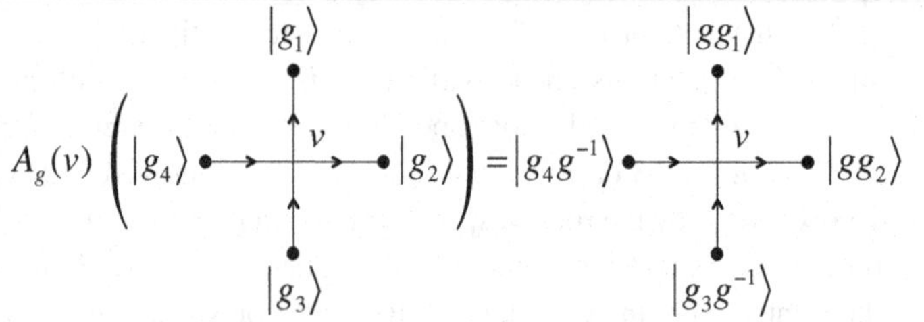

Fig. 5.14 A pictorial representation of the action of $A_g(v)$ on the link states of a vertex v.

This assigns energy to states that violate the stabiliser conditions and thus protects the code from undesired perturbations.

Similarly to (5.37) we can define projectors onto quasiparticle occupations of the vertices. There is a large variety of particles associated with the vertices and plaquettes of $D(S_3)$ (Bais *et al.*, 1992). Here we shall be concerned with only the vertex quasiparticles, Λ and Φ, defined through the projectors

$$P_\Lambda(v) = \frac{1}{6}\left[A_e(v) + A_c(v) + A_{c^2}(v) - A_t(v) - A_{tc}(v) - A_{tc^2}(v)\right], \qquad (5.40)$$

$$P_\Phi(v) = \frac{1}{3}\left[2A_e(v) - A_c(v) - A_{c^2}(v)\right]. \qquad (5.41)$$

Projectors (5.38), (5.40), (5.41) are all orthogonal to each other, i.e., $P_X(v)P_{X'}(v) = 0$ for $X \neq X'$. Importantly, $P_\Lambda(v)$ defines through the condition $P_\Lambda|\psi\rangle = |\psi\rangle$ the occupation of quasiparticle Λ at v and, similarly, $P_\Phi(v)$ defines quasiparticle Φ. Quasiparticles Λ and Φ are created from the stabiliser space by acting on a single spin s with the following operators:

$$W_\Lambda(s) = |e\rangle\langle e| + |c\rangle\langle c| + \left|c^2\right\rangle\left\langle c^2\right| - |t\rangle\langle t| - |tc\rangle\langle tc| - \left|tc^2\right\rangle\left\langle tc^2\right|, \quad (5.42)$$

$$W_\Phi(s) = 2|e\rangle\langle e| - |c\rangle\langle c| - \left|c^2\right\rangle\left\langle c^2\right|. \qquad (5.43)$$

Direct application of the projectors $P_\Lambda(v)$ and $P_\Phi(v)$ shows that W_Λ and W_Φ create anyons on the two vertices that share the rotated spin. A protocol to move anyons several edges apart is given in Aguado *et al.* (2008).

Having established the explicit form of the creation operators of the anyons helps us to identify their properties. All of the particles, 1, Λ and Φ, have trivial statistics with respect to each other, but they can still exhibit non-Abelian behaviour. When anyons of different type are brought to the same vertex, the possible outcomes are given by the fusion rules

$$\Lambda \times \Lambda = 1, \quad \Lambda \times \Phi = \Phi, \quad \Phi \times \Phi = 1 + \Lambda + \Phi. \qquad (5.44)$$

In other words, two Λ anyons fuse to the vacuum, while a Λ can be absorbed in a Φ without changing its nature. The last fusion rule implies that Φ's are non-Abelian anyons that have three possible fusion channels 1, Λ and Φ. Note that even if the $D(S_3)$ model supports more particles, these fusion relations are closed, involving only 1, Λ and Φ.

One can verify these fusion rules from the creation operators (5.42) and (5.43) of the anyons. Consider a certain spin s. Direct application of (5.42) and (5.43) shows that

$$W_\Lambda(s)W_\Phi(s) = W_\Phi(s), \qquad (5.45)$$

which demonstrates that creating a Λ and a Φ anyon on the same vertex is equivalent to having only a Φ anyon on that vertex. Let us now place two Φ's on the same vertex. By direct algebra one can show that

$$W_\Phi(s)W_\Phi(s) = 4|e\rangle\langle e| + |c\rangle\langle c| + \left|c^2\right\rangle\left\langle c^2\right| = W_1(s) + W_\Lambda(s) + W_\Phi(s). \qquad (5.46)$$

Hence, the fusion of two Φ's can in general result in all types of particles, 1, Λ and Φ. In the following we shall see how these properties can be employed for encoding qubits with non-Abelian anyons.

5.5 Quantum doubles as quantum memories

It is our aim now to employ the $D(S_3)$ non-Abelian quantum double model, presented in Section 5.4, as a quantum memory. There exist many explicit strategies in the literature on how to employ Abelian anyons for this purpose (Dennis *et al.*, 2002; Raussendorf *et al.*, 2007). The non-Abelian cases seem to be treated on a more abstract level (Bravyi, 2006; Georgiev, 2006). Compared to Abelian anyons, such as the toric code, non-Abelian anyons have certain advantages in encoding and manipulating information. For example, they do not employ the genus of the surface to encode qubits, but rather the number of anyons on a surface with trivial topology.

5.5.1 Non-Abelian information encoding and manipulation

As shown in Section 5.4, the non-Abelian $D(S_3)$ model offers a simple subset of particles, namely the charges 1, Λ and Φ that satisfy the fusion rules

$$\Lambda \times \Lambda = 1, \quad \Lambda \times \Phi = \Phi, \quad \Phi \times \Phi = 1 + \Lambda + \Phi. \tag{5.47}$$

We employ the last fusion rule to encode the qubit states in the fusion outcomes 1 and Λ of two Φ's.

To encode a qubit consider four neighbouring vertices, as shown in Figure 5.15. Applying W_Φ to spins 1 and 3 on the vacuum state $|\xi\rangle$ creates two pairs of Φ charges: one on vertices v_1 and v_2 and the other on v_3 and v_4. Since these pairs are created from the vacuum, they both carry the vacuum fusion channel when they are fused horizontally. This state is identified with the logical qubit state $|0_L\rangle$. Applying the operator W_Λ to spin 4 of the resulting state creates a pair of Λ particles. One of these will fuse with the Φ on v_1

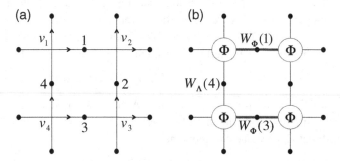

Fig. 5.15 (a) Four vertices, v_1, v_2, v_3 and v_4, are considered as the minimal system to encode one qubit by employing four Φ anyons. (b) Operations $W_\Phi(1)$ and $W_\Phi(3)$ generate two pairs of Φ anyons, each one with a vacuum fusion channel, i.e., qubit state $|0_L\rangle$. Application of $W_\Lambda(4)$ transforms both pairs into the Λ fusion channel, i.e., qubit state $|1_L\rangle$.

and the other with the Φ on v_4. After these fusions, the two horizontal Φ pairs carry the Λ fusion channel. This state is identified with the logical qubit state $|1_L\rangle$. The same is achieved if W_Λ is applied to spin 2. The encoded states for the logical qubit can be written explicitly as

$$|0_L\rangle = W_\Phi(1)W_\Phi(3)|\xi\rangle,$$
$$|1_L\rangle = W_\Lambda(4)W_\Phi(1)W_\Phi(3)|\xi\rangle, \tag{5.48}$$

where $|\xi\rangle$ is the vacuum state. Both logical states are four Φ anyon states, but with different pairwise fusion channels.

It is possible to move the encoding Φ anyons apart rather than keeping them all on neighbouring vertices. Then their fusion outcome, and hence the encoded information, is topologically protected from errors that act as local perturbations to the Hamiltonian. On the other hand, an error in the form of a spurious Λ anyon present between two Φ's can be fused with the Φ anyon that is closest to it. This error correcting step is similar to the one we employed in the toric code. Here the distance of the code is the geometric distance between the Φ anyons that encode the information.

In (5.48) the operators act on a single spin, s, and they generate neighbouring anyons. When the anyons are positioned further apart, the operations act on a corresponding chain of spins, C, lying on a path connecting the anyons. The operations $W_\Lambda(C)$ and $W_\Phi(C)$ that create these anyons at distant locations take the form

$$W_\Lambda(C) = \prod_{s \in C} W_\Lambda(s),$$
$$W_\Phi(C) = \sum_{k=0,1,2} \sum_{g_n \times \ldots \times g_1 = c^k} (\omega^k + \omega^{-k})|g_1, \ldots, g_n\rangle \langle g_1, \ldots, g_n|. \tag{5.49}$$

In this definition $g_1, \ldots, g_n$ are the states of the spins within the chain C, c is the element of S_3, while $\omega = e^{i2\pi/3}$. When $n = 1$ these definitions reduce to (5.42) and (5.43). The logical states can be written as in (5.48). By employing $4n$ anyons of type Φ we can therefore encode n qubits.

After encoding a qubit in two pairs of Φ anyons we would like to perform logical operations on it. A logical X operation corresponds to a process that creates two Λ charges and fuses both with a Φ from each pair. This evolution is described by

$$X = W_\Lambda(C), \tag{5.50}$$

where C is a path that connects two Φ's, one from each pair. Note that $W_\Lambda(C)$ does not correspond to braiding, so strictly speaking it is not a topological gate.

The logical Z operation corresponds to vertex operators acting on both Φ charges of either pair. For example, it can be given by

$$Z = A_t(v_1)A_t(v_2). \tag{5.51}$$

Such an operation corresponds to the topological evolution of taking a flux anyon that resides on plaquettes and braiding it around v_1 and v_2. The further apart we keep the vertices v_1 and v_2 the larger the loop the flux anyon needs to make in order to perform a

Z operation. Hence, the encoded information is better protected from stray flux anyons when the encoding anyons are kept far apart. At the same time, the difficulty of the control procedure for performing the Z gate increases. This trade-off between the degree of protection against local errors and accessibility of the encoded information is generic in topological models. What topological models offer us is the means to outperform environmental errors in accessing the encoded information.

Since the logical qubit is encoded in the fusion channel, measurement in the Z basis is achieved through fusion of charge pairs. Fusing the Φ charges on v_1 and v_2 and obtaining the vacuum or a Λ implies a logical qubit state of $|0_L\rangle$ or $|1_L\rangle$, respectively. Introducing non-topological spin operations can make this simple scheme universal (Wootton *et al.*, 2009). By employing flux and charge anyons in the encoding of qubits and braiding operations, it is possible to devise a more complicated scheme for universal quantum computation with purely topological means (Aguado *et al.*, 2008; Mochon, 2004).

Summary

In this chapter we presented the relation between quantum error correcting codes and a specific family of topological spin lattice models called quantum double models. Having an anyonic model described in terms of spin states enables us to study the spin rotations that are necessary to manipulate the states of the anyons. Abelian anyon models, such as the toric code, serve well in protecting quantum information from a variety of errors. However, these anyons have simple mutual statistics, meaning that information processing by purely topological means is not possible. To achieve universal quantum computation one can additionally employ non-topological operations, such as spin measurements (Wootton *et al.*, 2009). Non-Abelian anyons have a richer structure and can support universal quantum computation. These models are rather complicated, as for example they need spin lattices with at least six level spins, posing challenges for their realisation in the laboratory. Inspiring proposals exist suggesting how we could implement these models with cold atom technology (Weimer *et al.*, 2010) or with Josephson junctions (Doucot *et al.*, 2004; Gladchenko *et al.*, 2009). Architectures have also been proposed that produce effective Hamiltonians for quantum doubles using two-spin interaction terms only (Brell *et al.*, 2011).

Having an encoding space of quantum states that is protected from errors is an important task for technological applications. Turning an error correcting code into a Hamiltonian combines a finite energy gap above the code space that penalises the generation of errors with the non-local behaviour of the code states. This is a generic property of the topological models that becomes explicit in the case of quantum doubles. We should bear in mind that only coherent errors can be suppressed by the gap, i.e., spurious perturbations to the Hamiltonian (Bravyi *et al.*, 2010). The latter can cause virtual excitations which are automatically suppressed by keeping the characteristic size of the system large compared to the length of the perturbation. If the errors are generated by thermal noise, then the mechanism described above cannot automatically correct them (Dennis *et al.*, 2002). To deal with this problem, alternative methods have to be considered that are the subject of ongoing research.

Exercises

5.1 Demonstrate explicitly that the $A(v)$ and $B(p)$ operators, defined in (5.21) and (5.22), respectively, commute with each other for any v and p.

5.2 Develop explicitly the quantum double theory for the Z_2 group. The $D(Z_2)$ model is the toric code.

5.3 Establish the fourfold degeneracy of the toric code by demonstrating that non-trivial loops with $C_{\sigma^z}^1$ and $C_{\sigma^z}^2$ operations or combinations of them with $C_{\sigma^x}^1$ and $C_{\sigma^x}^2$ give states that linearly depend on the states (5.16).

5.4 What is the ground state degeneracy of the toric code model, defined on an infinite plane with an even number of punctures in the form of absent vertex or plaquette interaction terms in the Hamiltonian?

In this chapter, we consider Kitaev's honeycomb lattice model (Kitaev, 2006). This is an analytically tractable spin model that gives rise to quasiparticles with Abelian as well as non-Abelian statistics. Some of its properties are similar to the fractional quantum Hall effect, which has been studied experimentally in great detail even though it evades exact analytical treatment (Moore and Read, 1991). Due to its simplicity, the honeycomb lattice model is likely to be the first topological spin model to be realised in the laboratory, c.g., with optical lattice technology (Micheli *et al.*, 2006). Understanding its properties can facilitate its physical realisation and can provide a useful insight into the mechanisms underlining topological insulators and the fractional quantum Hall effect.

The honeycomb lattice model comprises interacting spin-1/2 particles arranged on the sites of a honeycomb lattice. It is remarkable that such a simple model can support a rich variety of topological behaviours. For certain values of its couplings, Abelian anyons emerge that behave like the toric code anyons. For another coupling regime, non-Abelian anyons emerge that correspond to the Ising anyons. The latter are manifested as vortex-like configurations of the original spin model that can effectively be described by Majorana fermions. These are fermionic fields that are antiparticles of themselves. They were first introduced in the context of high-energy physics (Majorana, 1937) and become increasingly important in the analysis of solid state phenomena (Wilczek, 2009). In this chapter we focus on the coupling regime of the honeycomb lattice that gives rise to non-Abelian anyons.

As Kitaev's honeycomb lattice model can support Ising anyons, we can employ it to perform topological quantum computation. Information encoded in the multiple fusion channels of the anyons remains protected from environmental perturbations as long as the anyons are kept far apart. We shall describe how one can transport the anyonic quasiparticles by local manipulations of the honeycomb lattice couplings. Hence, we can readily braid the anyons, thereby evolving the encoded information. Nevertheless, we have already seen that this model cannot support universal quantum computation by topological operations alone. Thankfully, the honeycomb lattice naturally exhibits short-range interactions between quasiparticles. The resulting non-topological evolutions can be used to supplement the braiding operations to comprise a universal set of gates (Bravyi, 2006; Das Sarma *et al.*, 2005).

Since its introduction in 2005, the honeycomb lattice model has been at the centre of numerous research efforts. The initial focus was on the Abelian phase as it provided a physically plausible way to obtain the toric code model (Kells *et al.*, 2008; Schmidt *et al.*, 2008; Vidal *et al.*, 2008). Subsequently, attention was given to the non-Abelian phase that supports Ising anyons. Several studies have been performed that relate the non-Abelian phase with the presence of edge states (Lee *et al.*, 2007), topological degeneracy (Kells *et al.*, 2009) and entanglement entropy (Chung *et al.*, 2010; Yao and Qi, 2010). Possible generalisations of this model have also been considered (Yao and Kivelson, 2007; Yao *et al.*, 2009). The presence of interactions between vortices was established in Lahtinen *et al.* (2008) and explicit demonstration of the non-Abelian statistics was obtained in Lahtinen and Pachos (2009).

To approach the honeycomb lattice model we employ two main elements, the energy spectrum and the corresponding energy eigenstates. By studying the spectrum for a variety of vortex configurations we identify the fusion degrees of freedom that correspond to non-Abelian anyons. To calculate the non-Abelian statistics of the vortices, we adiabatically transport them around each other. The corresponding evolution is given by a non-Abelian geometric phase, which coincides with the Ising statistics.

6.1 Introducing the honeycomb lattice model

In our analysis of the honeycomb lattice model we follow the fermionisation approach introduced by Kitaev (Kitaev, 2006; Pachos, 2007). This allows us to rewrite the spin model in terms of Majorana fermions, which are subject to an effective gauge field. For simplicity, we restrict ourselves to the parametric regime that supports non-Abelian anyons. These anyons emerge as the vortex-like quasiparticles of the gauge field. The methods presented in this section are applicable to models beyond Kitaev's honeycomb lattice (Hamma *et al.*, 2005; Kargarian and Fiete, 2010; Kells *et al.*, 2010; Yao and Kivelson, 2007).

6.1.1 The spin lattice Hamiltonian

Kitaev's honeycomb lattice model (Kitaev, 2006) consists of spin-1/2 particles residing at the sites of an infinite honeycomb lattice, as shown in Figure 6.1. We assign labels x, y and z to all links aligned along the same direction. We also bi-colour the lattice such that each black site is only connected to white ones and vice versa. The colouring reveals two triangular sub-lattices that comprise the honeycomb lattice. The spins interact according to a Hamiltonian

$$H = -J_x \sum_{x \text{ links}} \sigma_i^x \sigma_j^x - J_y \sum_{y \text{ links}} \sigma_i^y \sigma_j^y - J_z \sum_{z \text{ links}} \sigma_i^z \sigma_j^z - K \sum_{(i,j,k)} \sigma_i^x \sigma_j^y \sigma_k^z, \quad (6.1)$$

where J_x, J_y and J_z characterise positive nearest-neighbour couplings. Different models emerge for different values of these couplings.

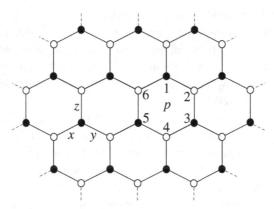

Fig. 6.1   The honeycomb lattice with spin-1/2 particles residing at its sites. This is a bi-colourable lattice. We label all links as x, y and z according to their orientation. A single plaquette p is depicted with its six sites enumerated.

In the dimerised limit, where one of the J couplings is larger than the sum of the other two, a topological phase is realised that supports the toric code. In the following we take the J couplings all equal, $J_x = J_y = J_z = J$, which corresponds to the non-Abelian topological phase (Kitaev, 2006). The last term in (6.1) is an effective magnetic field with coupling K. It can emerge as a perturbation when we introduce a Zeeman term (Jiang *et al.*, 2011; Kitaev, 2006), in which case $|K| \ll J$. Here we leave K as a freely tunable parameter. The sum runs over the sites such that every plaquette p contributes the six terms

$$\sum_{(i,j,k)\in p} \sigma_i^x \sigma_j^y \sigma_k^z = \sigma_1^z \sigma_2^y \sigma_3^x + \sigma_2^x \sigma_3^z \sigma_4^y + \sigma_3^y \sigma_4^x \sigma_5^z + \sigma_4^z \sigma_5^y \sigma_6^x + \sigma_5^x \sigma_6^z \sigma_1^y + \sigma_6^y \sigma_1^x \sigma_2^z,$$

where the enumeration around the plaquette p is given in Figure 6.1. The physical motivation to add this term is that it explicitly breaks time-reversal invariance, while preserving the exact solvability of the model. To be more precise, time-reversal symmetry is described by an anti-linear unitary operator $\hat{T}$, which acts on Pauli operators as

$$\hat{T} \sigma_i^\alpha \hat{T}^\dagger = -\sigma_i^\alpha. \tag{6.2}$$

Any product of an even number of Pauli operators with real coefficients will therefore respect the time-reversal symmetry. On the other hand, any odd product, such as the coupling of the spins with an external magnetic field or the three-spin term given in (6.1), will violate it. As we shall see below, violation of the time-reversal symmetry allows for non-trivial topological behaviour in the system as in the fractional quantum Hall effect.

The advantage of using a Hamiltonian with a three-spin coupling is that it has a local symmetry. Consider the plaquette operators

$$\hat{w}_p = \sigma_1^x \sigma_2^y \sigma_3^z \sigma_4^x \sigma_5^y \sigma_6^z. \tag{6.3}$$

These Hermitian operators square to the identity,

$$\hat{w}_p^2 = \mathbb{1}, \tag{6.4}$$

so their eigenvalues are $w_p = \pm 1$. Moreover, they commute with each other,

$$[\hat{w}_p, \hat{w}_{p'}] = 0 \text{ for all } p, \ p', \tag{6.5}$$

and with the Hamiltonian,

$$[H, \hat{w}_p] = 0 \text{ for all } p. \tag{6.6}$$

Relation (6.6) defines local symmetries that are at the heart of the solvability of the model.

Since the $\hat{w}_p$ are conserved quantities, the Hilbert space $\mathcal{L}$ of N spins on a plane with open boundaries can be partitioned into $2^{N/2}$ sectors $\mathcal{L}_w$ of dimension $2^{N/2}$. Each sector $\mathcal{L}_w$ is labelled by a distinct pattern $w = \{w_p\}$ of the eigenvalues $w_p = \pm 1$. Hence, the Hamiltonian can be reduced to each sector and the corresponding physics can be considered independently. This significantly reduces the complexity of the problem.

6.1.2 Majorana fermionisation

Our goal is to bring the Hamiltonian (6.1) in quadratic form by representing the spin operators with Majorana fermions (Kitaev, 2006). The quadratic Hamiltonian can then be directly diagonalised as it describes non-interacting particles. We start by considering a given site i and we introduce two fermionic modes $a_{1,i}$ and $a_{2,i}$ for each spin-1/2 particle, as depicted in Figure 6.2. Let us adopt a decomposition of the fermionic modes $a_{1,i}$ and $a_{2,i}$ into their 'real' and 'imaginary' parts as

$$c_i = a_{1,i} + a_{1,i}^\dagger, \quad b_i^x = i(a_{1,i}^\dagger - a_{1,i}), \quad b_i^y = a_{2,i} + a_{2,i}^\dagger, \quad b_i^z = i(a_{2,i}^\dagger - a_{2,i}). \tag{6.7}$$

The motivation for the distinction between b's and c's will become apparent in the following. The operators c_i, b_i^x, b_i^y and b_i^z are anticommuting and fermionic. In addition they satisfy the 'reality' condition

$$b_i^{\alpha\dagger} = b_i^\alpha, \quad c_i^\dagger = c_i, \tag{6.8}$$

i.e., they are their own antiparticles. This is the defining relation of Majorana fermions. Here, the Majorana operators are introduced as a tool to solve the Hamiltonian. Later, they appear again in a somewhat different formulation to describe the properties of the honeycomb lattice model and its emerging quasiparticles.

We would now like to represent the spin operators in terms of Majorana fermions. There is a redundancy in the fermionic encoding of the spins. Each spin-1/2 particle has a two-dimensional space, while the two 'complex' fermions, or the four Majorana fermions, have

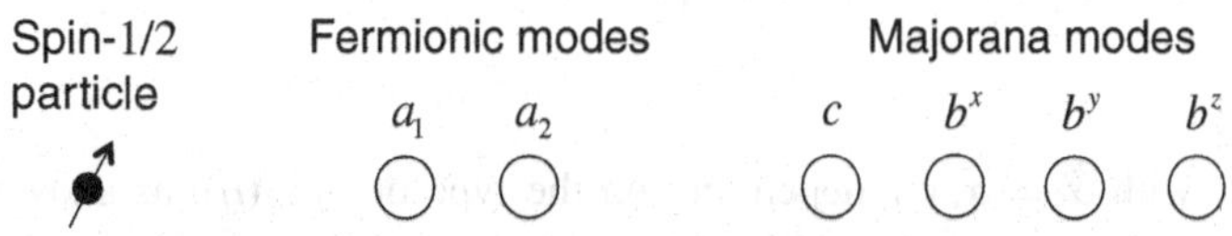

Fig. 6.2 A single spin-1/2 particle at a site of the lattice is described by two fermionic modes a_1 and a_2. Each of the two fermionic modes are decomposed in two sets of Majorana modes, c, b^x and b^y, b^z, respectively.

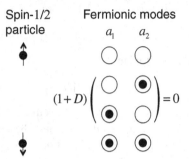

Fig. 6.3 Having both fermionic modes, a_1 and a_2, empty corresponds to the spin up state, while when they are both occupied they correspond to spin down. Single occupancies are projected out of the physical Hilbert space.

a four-dimensional space. To make the mapping consistent we need to project out two of the fermionic states. As illustrated in Figure 6.3 we choose to represent spin up having both fermionic modes empty and spin down having both modes occupied. This means

$$|\uparrow\rangle = |00\rangle, \quad |\downarrow\rangle = |11\rangle \tag{6.9}$$

with $a_1|00\rangle = a_2|00\rangle = 0$ and $|11\rangle = a_1^\dagger a_2^\dagger|00\rangle$. This representation is faithful if we restrict to the subspace $\mathcal{L}$ of fermionic states $|\Psi\rangle$ that satisfy

$$D_i|\Psi\rangle = |\Psi\rangle, \tag{6.10}$$

where

$$D_i = (1 - 2a_{1,i}^\dagger a_{1,i})(1 - 2a_{2,i}^\dagger a_{2,i}) = b_i^x b_i^y b_i^z c_i. \tag{6.11}$$

In this subspace we can make the following identification:

$$\sigma_i^\alpha = ib_i^\alpha c_i \text{ for } \alpha = x, y, z, \tag{6.12}$$

which satisfies

$$[D_i, \sigma_j^\alpha] = 0 \text{ and } i\sigma_i^x \sigma_i^y \sigma_i^z = b_i^x b_i^y b_i^z c_i = \mathbb{1}. \tag{6.13}$$

This representation satisfies the algebra of the Pauli matrices, $[\sigma^\alpha, \sigma^\beta] = i\varepsilon^{\alpha\beta\gamma}\sigma^\gamma$, only when we restrict to states that belong to $\mathcal{L}$.

Employing the spin representation $\sigma_i^\alpha = ib_i^\alpha c_i$ the Hamiltonian interactions become

$$\sigma_i^\alpha \sigma_j^\alpha = -i\hat{u}_{ij} c_i c_j \quad \text{and} \quad \sigma_i^x \sigma_j^y \sigma_k^z = -i\hat{u}_{ik}\hat{u}_{jk} D_k c_i c_j, \tag{6.14}$$

where we define the link operators

$$\hat{u}_{ij} = ib_i^\alpha b_j^\alpha, \tag{6.15}$$

with $\alpha = x, y, z$ depending on the type of link (ij), as shown in Figure 6.1. The $\hat{u}_{ij}$ are antisymmetric Hermitian operators that satisfy

$$\hat{u}_{ij} = -\hat{u}_{ji}, \quad \hat{u}_{ij}^2 = 1, \quad \hat{u}_{ij}^\dagger = \hat{u}_{ij}. \tag{6.16}$$

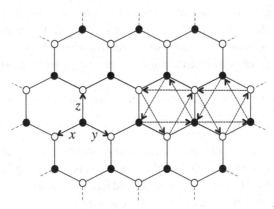

Fig. 6.4 The orientation of the x, y and z links is from the black sites towards the white ones. The next-to-nearest neighbour interactions are also depicted with their orientations.

Employing (6.14) and restricting the states of the system to the physical space $\mathcal{L}$, the Hamiltonian (6.1) takes the form

$$H = \frac{i}{4} \sum_{i,j} \hat{A}_{ij} c_i c_j, \qquad \hat{A}_{ij} = 2J_{ij}\hat{u}_{ij} + 2K \sum_{k} \hat{u}_{ik}\hat{u}_{jk}. \tag{6.17}$$

The $\hat{A}_{ij}$ are link operators, while the c_i's are independent Majorana modes that reside on the sites. The first term in $\hat{A}_{ij}$ corresponds to nearest-neighbour interactions between the c_i's. The second term describes next-to-nearest neighbour interactions between sites i and j that are linked through site k. To consistently define the antisymmetric operator $\hat{u}_{ij}$ throughout the lattice we need to define an orientation of the links, such as the one given in Figure 6.4. We assign an overall $+$ sign to every term that involves sites i and j when the arrow points from i to j and a $-$ sign when the arrow points from j to i. If two sites are not connected by an arrow the corresponding $\hat{A}_{ij}$ element is zero.

6.1.3 Emerging lattice gauge theory

The fermionisation of the spin degrees of freedom provides a new physical interpretation of the honeycomb lattice model. To reveal this let us consider the properties of Hamiltonian (6.17). It can be verified that

$$[H, D_i] = 0, \tag{6.18}$$

where H is now written in terms of the Majorana operators. Diagonalising the Majorana Hamiltonian is hence compatible with restricting the possible states of the system to the physical subspace $\mathcal{L}$ with eigenvalue $+1$ for all the D_i operators. This condition can be interpreted as a Gauss law. States that belong to the spectrum of the original spin system need to be eigenstates of Hamiltonian (6.17) as well as satisfy the Gauss law constraint, i.e.,

$$H \,|\, \Psi \rangle = E \,|\, \Psi \rangle \ \text{and} \ D_i \,|\, \Psi \rangle = |\, \Psi \rangle. \tag{6.19}$$

There are two equivalent ways of finding the solution to these equations. Either we start with the states that satisfy the Gauss law (i.e., belonging in the symmetric space $\mathcal{L}$) and then construct eigenstates of the Hamiltonian within $\mathcal{L}$ or we first find all eigenstates of the Hamiltonian and then symmetrise them in order to satisfy the Gauss law. In the following we take the latter approach.

In the model considered here, the link operators $\hat{u}_{ij}$ are local symmetries, i.e.,

$$[H, \hat{u}_{ij}] = 0. \tag{6.20}$$

Hence, one could imagine assigning patterns of eigenvalues u_{ij} to the oriented links of the lattice. Doing so, the Hamiltonian (6.17) becomes quadratic in c's and can be diagonalised directly. However, operators D_i and $\hat{u}_{ij}$ do not share a common set of eigenstates as they do not commute. In fact, they anticommute:

$$\{\hat{u}_{ij}, D_i\} = 0. \tag{6.21}$$

Sectors labelled by the eigenvalue patterns $u = \{u_{ij} = \pm 1\}$ are hence not part of $\mathcal{L}$. In other words, solving Hamiltonian (6.17) for a given pattern of u_{ij}'s results in eigenstates that do not necessarily satisfy constraint (6.10). On the other hand, the plaquette operators $\hat{w}_p$ become the products of the link operators

$$\hat{w}_p = \prod_{i,j \in p} \hat{u}_{ij}. \tag{6.22}$$

Importantly, they still commute with the Hamiltonian and with the D_i operators:

$$[\hat{w}_p, H] = 0 \quad \text{and} \quad [\hat{w}_p, D_i] = 0. \tag{6.23}$$

Eigenstates of the plaquette operators can hence belong in the physical subspace $\mathcal{L}$.

The above observation allows for the following lattice gauge theory interpretation. Consider Figure 6.5. The eigenvalues $\{u_{ij} = \pm 1\}$ of the link operators can be thought of as a classical Z_2 gauge field. Taking this approach, the operators D_i perform local gauge

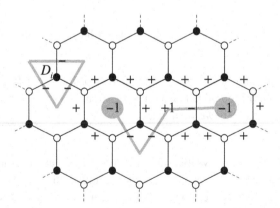

Fig. 6.5 The application of a D_i operator at a site changes by a $-$ sign the u_{ij} link eigenvalues around an elementary loop. A given pattern of eigenvalues $u_{ij} = \pm 1$ of the link operators gives a pattern of vortices in terms of eigenvalues $w_p = \pm 1$ of the plaquette operators. Two vortices are depicted as grey disks where the plaquette eigenvalue is -1. All link eigenvalues are taken to be $+1$ except for the ones along a string connecting the vortices with values -1.

transformations on them. Indeed, applying D_i to an eigenstate of all $\hat{u}_{ij}$ inverts the sign of the eigenvalues of the link operators that are connected with the site i. Such an operation on an initial $u_{ij} = +1$ configuration of the oriented link operators is illustrated in Figure 6.5. The application of a D_i keeps the states in the same vortex sector. Moreover, the gauge-invariant plaquette operators $\hat{w}_p$ can be identified with the Wilson loop operators. The eigenvalues $w_p = -1$ can hence be interpreted as having a π-flux vortex living on plaquette p. The different physical sectors of the model are then equivalent to configurations of vortices created by fixing the gauge u, i.e., by the pattern of the eigenvalues of the gauge field. There are many different patterns of u's that correspond to the same w configuration, so the function $w(u)$ is a many-to-one function. The pattern of eigenvalues $u_{ij} = -1$ can be visualised as an unphysical string passing through the link (ij) that either connects two vortices or belongs to a loop, as shown in Figure 6.5. The unphysicality follows from the violation of constraint (6.10) by the link operator eigenstates $| \Psi_u \rangle$ that belong in a given gauge sector u. To rectify this we perform the projection

$$| \Psi_w \rangle = \mathbf{D} \, | \Psi_u \rangle \,, \qquad \mathbf{D} = \prod_{i=1}^{N} \left(\frac{\mathbb{1} + D_i}{2} \right). \qquad (6.24)$$

This projection produces normalised physical states that correspond to the vortex configuration given by the pattern of w_p's. We shall call such a configuration the vortex sector. Due to the anticommutation of D_i and $\hat{u}_{ij}$, the physical state $| \Psi_w \rangle \in \mathcal{L}$ is an equal amplitude superposition of all loops and strings compatible with the vortex sector w. It is worth emphasising that the energy eigenvalues of the Hamiltonian are not affected by the symmetrisation of the states with the operator $\mathbf{D}$. Moreover, the properties we shall be focusing on in the following do not need its application.

Our starting point in this section was the honeycomb lattice model with spins residing at its sites, as indicated in Figure 6.6(a). The Majorana fermionisation introduced four Majorana fermions, c, b^x, b^y and b^z, at each site of the lattice, as shown in Figure 6.6(b). Fixing the eigenvalues $u_{ij} = \pm 1$ for each link removes the b operators completely from the model. This step reduces the initial spin model to a honeycomb lattice model of tunnelling Majorana fermions, as shown in Figure 6.6(c). In each sector parameterised by a certain pattern of u_{ij}'s, the model is equivalent to a problem of free Majorana fermions, whose

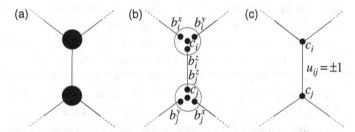

Fig. 6.6 (a) Two sites of the honeycomb lattice with spins residing on them. (b) The Majorana representation of the spins introduces four Majorana fermions for each spin, c, b^x, b^y and b^z. Link operators $\hat{u}_{ij} = ib_i^\alpha b_j^\alpha$ correspond at each link (ij). (c) Substituting the desired eigenvalue ± 1 of the link operators $\hat{u}_{ij}$ reduces the model to a honeycomb lattice of tunnelling Majorana fermions c.

tunnelling couplings depend on the underlying vortex configuration $w(u)$. Their Hamiltonian can hence be written as

$$H = \frac{i}{4} \sum_{i,j} A_{ij} c_i c_j, \tag{6.25}$$

where the $A_{ij}^{w(u)}$,

$$A_{ij}^{w(u)} = 2J_{ij} u_{ij} + 2K \sum_k u_{ik} u_{jk}, \tag{6.26}$$

are now a set of real numbers. This Hamiltonian is much simpler to solve than the initial one, which is given in terms of spins. To find its eigenvalues only requires diagonalising the matrix A_{ij}. This matrix grows polynomially with the linear size of the system instead of growing exponentially as the spin Hamiltonian does. The reason for this is that Hamiltonian (6.25) describes non-interacting particles hopping along the sites of the honeycomb lattice with tunnelling couplings given by A_{ij} in (6.26). Nevertheless, an exponential number of matrices A_{ij} needs to be diagonalised to find the spectrum of the complete model, one for each pattern of u eigenvalues. This will not be necessary as most of the physically relevant questions that we address in the following can be answered using specific gauge configurations.

6.2 Solving the honeycomb lattice model

In the previous section we employed the Majorana fermionisation method to simplify the honeycomb lattice Hamiltonian. It allowed us to represent the two-spin and three-spin interactions in terms of free Majorana fermions. These fermions were shown to be hopping on the honeycomb lattice between nearest and next-to-nearest sites with tunnelling couplings that depend on the u_{ij} eigenvalues.

In this section we employ two different approaches in studying the honeycomb lattice model. Initially, we take the continuum limit of the lattice Hamiltonian. This probes states with small energies that correspond to wavelengths much larger than the lattice spacing of the system. The resulting continuous Hamiltonian has a simple form. It corresponds to a relativistic quantum field theory that reveals the topological properties of the model. This methodology is also employed in the study of several other systems such as graphene (Wallace, 1947) and topological insulators (Hasan and Kane, 2010). Alternatively, one can perform direct diagonalisation of the Hamiltonian. This approach offers the possibility to obtain exact solutions for relatively large system sizes. We shall consider a variety of vortex configurations aiming to quantitatively study their anyonic character, such as their fusion and braiding properties.

6.2.1 The no-vortex sector

First, let us consider the honeycomb lattice model in its Majorana form. As A_{ij} given in (6.26) can be split into two terms, the Hamiltonian (6.25) can be written as

$$H = H_1 + H_2, \tag{6.27}$$

where the nearest-neighbour term is given by

$$H_1 = \frac{i}{4} \sum_{ij} 2J u_{ij} c_i c_j \tag{6.28}$$

and the next-to-nearest term is given by

$$H_2 = \frac{i}{4} \sum_{ij} 2K \sum_k u_{ik} u_{jk} c_i c_j. \tag{6.29}$$

The nearest-neighbour hopping amplitudes are taken to be uniform and equal to J. Moreover, we focus on the case where all u_{ij} along the orientation of Figure 6.4 are $+1$. This case corresponds to the absence of any vortices. Due to a theorem by Lieb (1994) it is known that the lowest energy states of the model belong to this vortex-free sector.

To proceed, we employ a Fourier transformation applied initially to H_1. The honeycomb lattice is bi-colourable, so the smallest unit cell for which it becomes periodic comprises two neighbouring sites. Let us take them to be the white and black sites along the z link, as shown in Figure 6.7. Here the Majorana modes are denoted by a and b, respectively. Employing

$$a_\mathbf{r} = \sum_\mathbf{p} e^{-i\mathbf{p}\cdot\mathbf{r}} a_\mathbf{p}, \tag{6.30}$$

and a similar relation for $b_\mathbf{r}$ and $c_\mathbf{r}$, we have that

$$\begin{aligned}
H_1 &= \frac{i}{4} 2J \sum_\mathbf{r} a_\mathbf{r}(b_{\mathbf{r}+\mathbf{s}_1} + b_{\mathbf{r}+\mathbf{s}_2} + b_{\mathbf{r}+\mathbf{s}_3}) + \text{h.c.} \\
&= \frac{i}{4} 2J \sum_{\mathbf{r},\mathbf{p},\mathbf{p}'} \Big(\sum_{\alpha=1,2,3} e^{-i\mathbf{p}\cdot\mathbf{r} - i\mathbf{p}\cdot(\mathbf{r}+\mathbf{s}_\alpha)} \Big) a_\mathbf{p} b_{\mathbf{p}'} + \text{h.c.} \\
&= \frac{i}{4} 2J \sum_{\mathbf{p},\mathbf{p}'} \Big(\sum_{\alpha=1,2,3} e^{-i\mathbf{p}\cdot\mathbf{s}_\alpha} \Big) a_{-\mathbf{p}} b_\mathbf{p} + \text{h.c.}
\end{aligned} \tag{6.31}$$

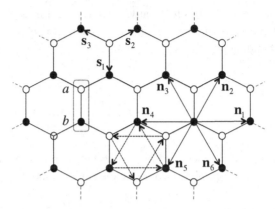

The unit cell of the honeycomb lattice includes two neighbouring sites a and b, taken here to be along a z link. The vectors $\mathbf{s}_1$, $\mathbf{s}_2$ and $\mathbf{s}_3$ are depicted, which define the three nearest neighbours on the lattice, as well as $\mathbf{n}_1, ..., \mathbf{n}_6$, which define the six next-to-nearest neighbours.

The vectors $\mathbf{s}_\alpha$ in this equation define the nearest neighbours of $a_\mathbf{r}$, as shown in Figure 6.7. To simplify notation, we define $\tilde{a} = e^{-i\pi/4}a$, $\tilde{b} = e^{i\pi/4}b$ and

$$f(\mathbf{p}) = 2J \sum_{\alpha=1,2,3} e^{-i\mathbf{p}\cdot\mathbf{s}_\alpha}. \tag{6.32}$$

Then the Hamiltonian H_1 takes the form

$$H_1 = \frac{1}{4}\sum_\mathbf{p} f(\mathbf{p})\tilde{a}_\mathbf{p}^\dagger \tilde{b}_\mathbf{p} + \text{h.c.} \tag{6.33}$$

The next-to-nearest neighbour interactions take the form

$$\begin{aligned}
H_2 &= \frac{iK}{2}\sum_{\mathbf{r},\mathbf{p},\mathbf{p}'} \left(e^{-i\mathbf{p}'\cdot\mathbf{n}_1} - e^{-i\mathbf{p}'\cdot\mathbf{n}_2} + e^{i\mathbf{p}'\cdot(\mathbf{n}_1-\mathbf{n}_2)} - e^{i\mathbf{p}'\cdot\mathbf{n}_1} + e^{i\mathbf{p}'\cdot\mathbf{n}_2} - e^{-i\mathbf{p}'\cdot(\mathbf{n}_1-\mathbf{n}_2)} \right) \\
&\quad \times e^{-i(\mathbf{p}+\mathbf{p}')\cdot\mathbf{r}} c_\mathbf{p} c_{\mathbf{p}'} \\
&= \frac{1}{4}\sum_\mathbf{p} \Delta(\mathbf{p})(\tilde{a}_\mathbf{p}^\dagger \tilde{a}_\mathbf{p} - \tilde{b}_\mathbf{p}^\dagger \tilde{b}_\mathbf{p}),
\end{aligned} \tag{6.34}$$

where we define

$$\Delta(\mathbf{p}) = 4K\big(-\sin\mathbf{p}\cdot\mathbf{n}_1 + \sin\mathbf{p}\cdot\mathbf{n}_2 + \sin\mathbf{p}\cdot(\mathbf{n}_1 - \mathbf{n}_2) \big). \tag{6.35}$$

The vectors $\mathbf{n}_\alpha$ denote the next-to-nearest neighbours of a certain site, as shown in Figure 6.7. Combining H_1 and H_2 we obtain

$$H = \frac{1}{4}\sum_\mathbf{p} (\tilde{a}_\mathbf{p}^\dagger \ \tilde{b}_\mathbf{p}^\dagger) \begin{pmatrix} \Delta(\mathbf{p}) & f(\mathbf{p}) \\ f(\mathbf{p})^* & -\Delta(\mathbf{p}) \end{pmatrix} \begin{pmatrix} \tilde{a}_\mathbf{p} \\ \tilde{b}_\mathbf{p} \end{pmatrix}. \tag{6.36}$$

Since this form of the Hamiltonian is rather simple, it is now straightforward to find the eigenvalue of the energy for Majorana fermions with certain momentum by direct diagonalisation of the one-particle Hamiltonian

$$H(\mathbf{p}) = \begin{pmatrix} \Delta(\mathbf{p}) & f(\mathbf{p}) \\ f(\mathbf{p})^* & -\Delta(\mathbf{p}) \end{pmatrix}. \tag{6.37}$$

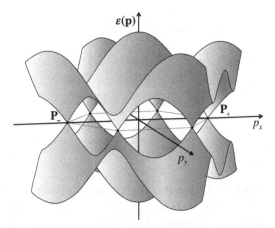

Fig. 6.8 The surface plots of the eigenvalues $\varepsilon(\mathbf{p}) = \pm|f(\mathbf{p})|$ are presented as a function of the momenta p_x and p_y. The upper surface corresponds to the positive sign and the lower surface to the negative sign. In the plotted contour there are two independent Fermi points $\mathbf{P}_+$ and $\mathbf{P}_-$. Note that in the case of a finite lattice system the momentum takes discrete values rather than the continuous ones shown here.

Suppose $K = 0$ so that $\Delta(\mathbf{p}) = 0$. Then, the two eigenvalues of $H(\mathbf{p})$ are given by

$$\varepsilon(\mathbf{p}) = \pm|f(\mathbf{p})| = \pm\frac{J}{2}\sqrt{1 + 4\cos^2\frac{\sqrt{3}p_x}{2} + 4\cos\frac{3p_y}{2}\cos\frac{\sqrt{3}p_x}{2}}. \qquad (6.38)$$

This relation is known as the dispersion relation of the energy as a function of the momentum. The positive energy is called the conductance band and the negative energy the valence band. Common as they might look, these energy eigenvalues possess a unique property. They become zero, $\varepsilon(\mathbf{p}) = 0$, for certain isolated values of momentum $\mathbf{p}$ known as the Fermi points. Two independent such momenta are given by

$$\mathbf{P}_\pm = \pm\left(\frac{4\pi}{3\sqrt{3}}, 0\right), \qquad (6.39)$$

as shown in Figure 6.8. The presence of Fermi points makes the low-energy behaviour of the honeycomb lattice model special. For example, it provides a powerful diagnostic tool for determining the properties of the model. Finally, it can be shown that the spectrum of Hamiltonian (6.36) consists of normal fermionic modes (Kitaev, 2000). In other words, the free particles of the theory are fermions. This property holds for any vortex configuration.

6.2.1.1 Continuous limit approximation

Hamiltonian (6.36) is an exact description of the infinite honeycomb lattice model without vortices. To proceed we employ the continuum limit approximation. What states of the model are well described by this method? Intuitively, eigenstates with small energy are characterised by large wavelengths. In the limit of infinitely large wavelength, the lattice spacing becomes negligibly small and the continuous picture can be employed safely. Moreover, the continuum limit method breaks down when high-energy states are considered where the lattice structure becomes energetically relevant.

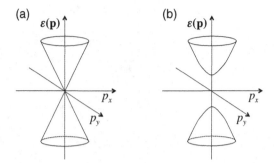

The dispersion relation $\varepsilon(\mathbf{p})$ near a single Fermi point as a function of small momenta. (a) The characteristic conical shape that emerges near the Fermi point, when $K = 0$ is depicted. It corresponds to a linear dispersion relation that characterises the Dirac equation. (b) A non-zero K gives rise to an energy gap between the valence and the conductance bands.

The ground state (i.e., the lowest energy state) is the state in which the valence band (negative fermionic states) is fully occupied. This state therefore corresponds to half filling. Small energy fluctuations around the ground state are described by the Hamiltonian (6.37) near its Fermi points. Fortunately, the low-energy regime, which determines the quantum phase of the system, is the regime of applicability for the continuum limit method.

The behaviour of $\varepsilon(\mathbf{p})$ for $K = 0$ and $\mathbf{p}$ near a Fermi point is given in Figure 6.9(a). This figure focuses on one of the Fermi points of Figure 6.8. Notably, the dispersion relation is linear with respect to the momentum, creating a cone similar to the behaviour of two-dimensional massless Dirac particles. Let us expand the Hamiltonian near the Fermi points by considering momenta $\mathbf{P}_{\pm} + \mathbf{p}$, where $\mathbf{p}$ is very small. Keeping only first order in the small parameters $\mathbf{p} = (p_x, p_y)$, we have

$$f(\mathbf{P}_{+} + \mathbf{p}) = -3J(p_x + ip_y) + \mathcal{O}(\mathbf{p}^2) \text{ and } f(\mathbf{P}_{-} + \mathbf{p}) = 3J(p_x - ip_y) + \mathcal{O}(\mathbf{p}^2). \quad (6.40)$$

Substituting these in (6.37) gives the Hamiltonian a Dirac-like form:

$$H_{+}(\mathbf{p}) = H(\mathbf{P}_{+} + \mathbf{p}) = 3J(-\sigma^x p_x + \sigma^y p_y) \quad (6.41)$$

and

$$H_{-}(\mathbf{p}) = H(\mathbf{P}_{-} + \mathbf{p}) = 3J(\sigma^x p_x + \sigma^y p_y), \quad (6.42)$$

respectively.

Let us now consider the case where K is different from zero. Then we have

$$\Delta(\mathbf{P}_{+} + \mathbf{p}) = 6\sqrt{3}K + \mathcal{O}(\mathbf{p}^2), \ \ \Delta(\mathbf{P}_{-}) = -\Delta(\mathbf{P}_{+}). \quad (6.43)$$

Interestingly, the K-term does not shift the position of the Fermi points as it does not contribute any linear momentum term. It only creates an energy gap

$$\Delta = 6\sqrt{3}K \quad (6.44)$$

between the valence and the conductance bands, as shown in Figure 6.9(b). At this point it is preferable to combine the two Hamiltonians, $H_{+}(\mathbf{p})$ and $H_{-}(\mathbf{p})$, together and treat the two Fermi points as two different pseudo-spin components. In particular, we consider the basis $\Psi(\mathbf{p}) = (\tilde{a}_+, \tilde{b}_+, \tilde{b}_-, \tilde{a}_-)^T$ so that the Hamiltonian becomes

$$H_{\text{tot}}(\mathbf{p}) = \begin{pmatrix} \Delta & -p & 0 & 0 \\ -\bar{p} & -\Delta & 0 & 0 \\ 0 & 0 & \Delta & p \\ 0 & 0 & \bar{p} & -\Delta \end{pmatrix} = -\sigma^z \otimes \sigma^x p_x + \sigma^z \otimes \sigma^y p_y + \mathbb{1} \otimes \sigma^z \Delta, \quad (6.45)$$

where $p = p_x + ip_y$ and $\bar{p} = p_x - ip_y$ and where we absorbed the factor $3J$ by rescaling the momentum. Since the matrices of the momentum terms satisfy the Clifford operator properties $\{\sigma^z \otimes \sigma^x, \sigma^z \otimes \sigma^y\} = 0$ and $(\sigma^z \otimes \sigma^x)^2 = (\sigma^z \otimes \sigma^y)^2 = \mathbb{1}$, this Hamiltonian is in the form of a Dirac operator with mass Δ.

The way the mass term appears in (6.45) gives the Hamiltonian some intriguing topological properties. To reveal them we verify that the matrices

$$\Sigma_1 = \sigma^z \otimes \sigma^x, \ \ \Sigma_2 = \sigma^z \otimes \sigma^y, \ \ \Sigma_3 = \mathbb{1} \otimes \sigma^z \quad (6.46)$$

of $H_{\text{tot}}(\mathbf{p})$ satisfy the su(2) algebra

$$[\Sigma_i, \Sigma_j] = 2i\epsilon_{ijk}\Sigma_k. \quad (6.47)$$

Hence, we can write

$$H_{\text{tot}}(\mathbf{p}) = \mathbf{\Sigma} \cdot \mathbf{n}(\mathbf{p}), \ \ \text{where } \mathbf{\Sigma} = (\Sigma_1, \Sigma_2, \Sigma_3) \text{ and } \mathbf{n}(\mathbf{p}) = (-p_x, p_y, \Delta). \quad (6.48)$$

The vector $\mathbf{n}$ not only parameterises the Hamiltonian, but also its eigenstates.

Consider the normalised vector $\hat{\mathbf{n}} = \mathbf{n}/|\mathbf{n}|$. This vector maps the torus to the unit sphere, as shown in Figure 6.10. Indeed, the momentum is periodic in both p_x and p_y directions, so it takes values on a torus. Moreover, all possible values of $\hat{\mathbf{n}}$ define a unit sphere. If $\Delta \neq 0$, then this map spans the whole unit sphere once when all the values of the momentum span the whole torus. Consider, for example, the case $\Delta > 0$. For momentum equal to $\mathbf{P}_+$ equations (6.43) and (6.45) imply that $\hat{\mathbf{n}} = (0, 0, 1)$ and the neighbourhood of $\mathbf{P}_+$ is mapped on the upper hemisphere. Similarly, at $\mathbf{P}_-$ we have $\hat{\mathbf{n}} = (0, 0, -1)$ and the neighbourhood of $\mathbf{P}_-$ is mapped on the lower hemisphere. The region between the two Fermi points is characterised by $|\mathbf{p}| \gg \Delta$ for which the vector $\hat{\mathbf{n}}$ is oriented along the equator, as can be seen from (6.48). In this case, the mapping of $\hat{\mathbf{n}}$ can be described by the Chern number (Avron *et al.*, 1983)

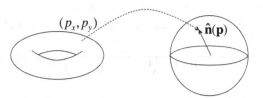

Fig. 6.10   The unit vector $\hat{\mathbf{n}}(\mathbf{p})$ defines a mapping from the torus to the unit sphere. The torus corresponds to the geometry of the periodic momentum coordinates, p_x and p_y, and the sphere corresponds to all possible orientations of $\hat{\mathbf{n}}(\mathbf{p})$. The Chern number (6.49) determines how many times $\hat{\mathbf{n}}(\mathbf{p})$ covers the whole sphere when the momentum spans the whole torus.

$$\nu = \frac{1}{4\pi} \int \int dp_x dp_y \, \frac{\partial \hat{\mathbf{n}}}{\partial p_x} \times \frac{\partial \hat{\mathbf{n}}}{\partial p_y} \cdot \hat{\mathbf{n}}, \tag{6.49}$$

which is equivalent to its previous definition (2.75). Here, the Chern number is the winding number of the map shown in Figure 6.10. The above arguments show that the unit vector defined from (6.48) gives

$$\nu = \frac{\Delta}{|\Delta|} = \text{sgn} \, \Delta = \pm 1. \tag{6.50}$$

In other words, this topological number returns the sign of Δ. As the mapping is between two surfaces, ν can be positive or negative corresponding to the relative orientations of the surfaces that are spanned. The non-zero value of ν for $K \neq 0$ signals that Hamiltonian (6.45) belongs to a topologically non-trivial class. As a consequence, the ground state has a non-local order, that is not encountered in usual insulators. Indeed, for non-trivial Chern numbers the vortices of the honeycomb lattice bind localised Majorana modes (Jackiw and Rossi, 1981; Kitaev, 2006). This is the second time we encounter Majorana fermionic modes. Now they appear as a collective effect that can be viewed as localised eigenstates of the Hamiltonian. In Section 6.3 it is shown that localised Majorana modes behave as Ising anyons. These general arguments illustrate that non-Abelian anyons can emerge from the honeycomb lattice model. A more direct study is presented in the following subsection.

6.2.2 Vortex sectors

The continuum approximation presented above allowed us to probe the no-vortex sector and understand its behaviour. It revealed the fermionic character of the spectrum and an energy gap that is present when $K \neq 0$. With this method we established the topological character of the system by evaluating a non-trivial Chern number. We now probe various configurations of vortices in order to understand their properties without using abstract arguments. Our aim is to find quantitative results for finite system sizes that can be treated numerically.

6.2.2.1 Finite lattice system

To study general vortex sectors, we define a finite lattice system with $2L_xL_y$ sites, where the couplings u can assume any desired pattern. The Hamiltonian (6.25) generalises to

$$H = \frac{1}{4} \left(\mathbf{c}_b^\dagger \; \mathbf{c}_w^\dagger \right) h \begin{pmatrix} \mathbf{c}_b \\ \mathbf{c}_w \end{pmatrix} \quad \text{with } h = \begin{pmatrix} h_{bb} & h_{bw} \\ h_{bw}^\dagger & -h_{bb}^T \end{pmatrix}. \tag{6.51}$$

Here, $\mathbf{c}_\lambda^\dagger = (c_{\lambda,1}^\dagger, \ldots, c_{\lambda,L_xL_y}^\dagger)$ for $\lambda = b, w$ and h_{bw} and h_{bb} are $L_xL_y \times L_xL_y$ matrices describing the nearest and next-to-nearest interactions, respectively (see Exercise 6.3). The $2L_xL_y \times 2L_xL_y$-dimensional matrix h is the corresponding one-particle Hamiltonian. The Hamiltonian has a double spectrum

$$h \left| \psi_i^\pm \right\rangle = \pm E_i \left| \psi_i^\pm \right\rangle, \quad \text{with } E_i \geq 0 \tag{6.52}$$

i.e., for every positive energy eigenstate, $\left| \psi_i^+ \right\rangle$, there is a corresponding negative one, $\left| \psi_i^- \right\rangle$. The fermionic nature of the Hamiltonian (6.51) dictates that the ground state energy, E_0, is given by the occupation of all the negative energy states. As these can only have a single fermionic occupancy we have

$$E_0 = -\sum_{i=1}^{L_x L_y} \frac{E_i}{2}, \qquad (6.53)$$

that can be evaluated when the spectrum (6.52) of h is known. Our aim is to keep the lattice size large in order to estimate the properties of the system in the thermodynamic limit.

6.2.2.2 Vortex manipulation and low-energy spectrum

In the previous section we demonstrated that when K is non-zero then there is a non-zero energy gap above the ground state. So a finite energy is needed in order to create a fermionic excitation. This gap is present in any vortex sector. Lieb's theorem (1994) also dictates that creating vortices costs a finite amount of energy. We denote by $2\Delta_v$ the gap that corresponds to the creation of a pair of vortices and by Δ the fermionic energy gap. As the vortices may interact, their gap is only defined in the asymptotic limit when all vortices are well separated and their interaction becomes negligible. To understand the relation between vortices and fermionic excitations we need to transport vortices and move between different vortex sectors.

From (6.22) we see that vortex configurations $w = \{w_p\}$ are created by arranging the link parameters $u = \{u_{ij}\}$. In order to manipulate w, one needs to locally manipulate u. This can be done effectively through the tunnelling coupling configurations J_{ij} between neighbouring sites (i, j) as well as K_{ijk} that corresponds to neighbouring sites (i, j, k). It can be seen from Hamiltonian (6.17) that the u_{ij} eigenvalues are paired with the local couplings J_{ij} and K_{ijk}. Therefore, the gauge configurations correspond to local configuration of these couplings. Manipulating these couplings instead of the pattern of u's gives the ability to transform the model in a continuous way. This provides an extension of the discrete transformations that are possible when we restrict to only changing u.

By manipulating the J and K couplings we can simulate the transport of vortices in a continuous way. Consider tuning the coupling configuration such that $J_z = -1$ on the d successive z links of the lattice, as shown in Figure 6.11. This amounts to creating two vortices separated linearly by d links. By varying d we can study the spectral evolution of the system as a function of the vortex separation. We can carry out the vortex transport 'continuously', if the sign of the coupling J_{ij} at the link $d + 1$ is reversed by taking $J_{ij} = 1 - 2s/S$ with $s \in [0, S]$. Moreover, if this protocol is carried out on a link between empty plaquettes or plaquettes with two vortices, the resulting process corresponds to the creation or annihilation of vortices, respectively. Under such tuning of couplings, the spectrum evolves smoothly between states belonging to different vortex sectors.

To obtain the low-energy spectrum of the total system, including vortices and fermions, we evaluate the fermionic spectrum of the Hamiltonian with zero and two vortices, numerically. The resulting energies are illustrated in Figure 6.12. We observe that in the absence of vortices the system is characterised by an energy gap, Δ, in agreement with (6.44). Such a gap is also there in the presence of vortices. But there are also fermionic modes that

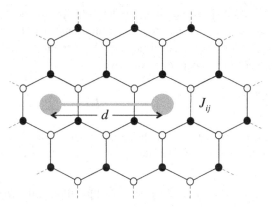

Fig. 6.11 When the oriented u's are taken $+1$ everywhere, two vortices at distance d apart can be represented as having $J = -1$ at the links crossed by the grey line. To smoothly transport the right vortex across a z link (ij) we parameterise its coupling as $J_{ij}(s) = 1 - 2s/S$ with $s \in [0, S]$.

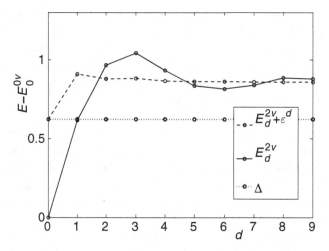

Fig. 6.12 The low-energy spectrum as a function of vortex separation d. The energies E in the two-vortex sector are with respect to the ground state energy E_0^{0v} of the vortex-free sector. The solid line is the total ground state energy, E_d^{2v}, the dashed one is the energy of the state with occupied zero mode, $E_d^{2v} + \varepsilon^d$, and the dotted line corresponds to the lowest-lying free fermion energy, Δ. For large separations d the energies E_d^{2v} and $E_d^{2v} + \varepsilon^d$ tend to the two-vortex energy $2\Delta_v$.

appear when we insert vortices with energies close to zero, i.e., below the energy gap. It appears that in the presence of $2n$ vortices, there are n such lowest-lying modes in the spectrum. These modes acquire asymptotically zero energy when the vortices are separated. In Figure 6.12 we see two states above the ground state that correspond to each zero mode being occupied or empty. Hence, the diagonalised Hamiltonian takes the form

$$H = \left[\sum_{i=n+1}^{L_x L_y} E_i b_i^\dagger b_i + \sum_{i=1}^{n} \varepsilon_i^d z_i^\dagger z_i - \left(\sum_{i=n+1}^{L_x L_y} \frac{E_i}{2} + \sum_{i=1}^{n} \frac{\varepsilon_i^d}{2} \right) \right], \qquad (6.54)$$

where b_i and z_i are fermionic annihilation operators (Lahtinen *et al.*, 2008). We have renamed the n smallest eigenvalues and the corresponding modes as ε_i^d and z_i, respectively.

Let us define the states that correspond to the low-energy modes of (6.54). Denote by $| \Psi_0 \rangle$ the ground state of the no-vortex sector and by $\left| \Psi_0^{nv} \right\rangle$ the ground state of a vortex sector with n vortices. Figure 6.12 shows the evolution of the lowest-lying states in the two-vortex sector relative to the ground state energy E_0^{0v} of the vortex-free sector. At large d, the two states $\left| \Psi_0^{2v} \right\rangle$ and $z_1^\dagger \left| \Psi_0^{2v} \right\rangle$ are degenerate with energy $2\Delta_v$ above the vortex-free ground state. They differ only by the occupation of the zero mode. As the vortices are brought closer, the degeneracy is lifted due to the mode $z_1^\dagger$ acquiring energy, i.e., ε_1^d becomes non-zero. As $d \to 0$, the vortices are brought to the same plaquette which corresponds to fusing them. It is indeed the interactions due to the microscopics of the model that give finite energy to the fusion modes, thereby making the two states $\left| \Psi_0^{2v} \right\rangle$ and $z_1^\dagger \left| \Psi_0^{2v} \right\rangle$ energetically distinguishable at small distances, d.

For $d \to 0$ we observe that the energy corresponding to $\left| \Psi_0^{2v} \right\rangle$ evolves to the energy of the ground state $| \Psi_0 \rangle$ of the vortex-free sector. On the other hand, $z_1^\dagger \left| \Psi_0^{2v} \right\rangle$ evolves to $b_{1,\mathbf{p}_0}^\dagger | \Psi_0 \rangle$, the first excited free fermion state in the vortex-free sector. Figure 6.12 also suggests that when the vortex pairs are far from each other, $\varepsilon_{i,\mathbf{p}_0}^d$ takes the form

$$\varepsilon_{1,\mathbf{p}_0}^d \sim \Delta^{2v} \cos(\dot{\omega} d) e^{-\frac{d}{\xi}}. \tag{6.55}$$

Here, $\omega > 0$ and $\xi > 0$ depend on the couplings J and K and parameterise the frequency of the oscillations and the convergence of the energy, respectively. As a conclusion, the small energy of these modes decays exponentially with the vortex separation and it has an oscillatory character (Cheng *et al.*, 2009; Lahtinen, 2011; Lahtinen *et al.*, 2008).

6.2.2.3 Zero modes as fusion degrees of freedom

The distinct behaviour of the two-vortex states $\left| \Psi_0^{2v} \right\rangle$ and $z_1^\dagger \left| \Psi_0^{2v} \right\rangle$ in Figure 6.12 as $d \to 0$ suggests that the occupation of the fermionic zero mode corresponds to the fusion channel of the vortices. This implies the identifications of the honeycomb lattice states with the Ising anyons given in Table 6.1. There, we identify the σ non-Abelian particles of the Ising anyon model with the vortices that are accompanied by zero modes. The ψ's correspond to occupied fermionic modes. Then, in accordance with the fusion rules (4.23), an occupied zero mode means that the σ's will fuse to a ψ, whereas an unoccupied mode implies that the fusion will give the vacuum 1. Note that, while the whole spectrum is diagonal in the basis of free fermions, the presence of two vortices splits a fermionic mode into two localised Majorana modes that sit in the vortex cores. How such Majorana fermions can give rise to Ising anyonic statistics is demonstrated in Section 6.3.

Due to fermionic parity conservation, there is no ground state degeneracy in the two-vortex case. Hence, one needs to consider at least the four-vortex case. The interpretation of Table 6.1 is further confirmed by the low-energy spectrum of such a four-vortex sector (Lahtinen *et al.*, 2008). When varying the separation of the two vortex pairs in a pairwise fashion, we obtain spectral lines similar to the ones given in Figure 6.12. Four Ising σ anyons have the physically non-trivial two-dimensional fusion spaces $\mathcal{M}_{(4)}$. The total

Table 6.1 Quasiparticles as Ising anyons		
Honeycomb lattice		Ising model
Ground state	$\leftrightarrow$	1, vacuum
Vortex	$\leftrightarrow$	σ, non-Abelian anyon
Fermionic excitation	$\leftrightarrow$	ψ, fermion

fusion channel of the four anyons can be either 1 or ψ. The spectral evolution shows that when vortices are fused, then there are two nearly degenerate states (either z_1 or z_2 occupied) that become the first excited state in the vortex-free sector. The states with neither or both zero modes occupied become the ground state or the two fermion state, respectively. Therefore, we can identify these four states with the fusion space basis states as

$$
\begin{aligned}
\left| \Psi_0^{2v} \right\rangle : &\qquad (\sigma \times \sigma)_1 \times (\sigma \times \sigma)_2 \to 1 \times 1 = 1, \\
z_1^\dagger z_2^\dagger \left| \Psi_0^{2v} \right\rangle : &\qquad (\sigma \times \sigma)_1 \times (\sigma \times \sigma)_2 \to \psi \times \psi = 1
\end{aligned}
\tag{6.56}
$$

and

$$
\begin{aligned}
z_1^\dagger \left| \Psi_0^{2v} \right\rangle : &\qquad (\sigma \times \sigma)_1 \times (\sigma \times \sigma)_2 \to \psi \times 1 = \psi, \\
z_2^\dagger \left| \Psi_0^{2v} \right\rangle : &\qquad (\sigma \times \sigma)_1 \times (\sigma \times \sigma)_2 \to 1 \times \psi = \psi.
\end{aligned}
\tag{6.57}
$$

Hence, the four-vortex fusion process is consistent with the Ising anyon description of the vortices.

6.2.2.4 Non-Abelian statistics of vortices

We now demonstrate the non-Abelian statistics of the vortices. For that we transport the vortices around each other in a continuous fashion and determine the corresponding evolution as a geometric phase acting on the fusion space. Similar calculations have been performed by using trial, instead of exact, wave functions for a variety of systems (Arovas *et al.*,1984; Baraban *et al.*, 2009; Tserkovnyak and Simon, 2003).

Consider four vortices created in pairs, one having the vacuum fusion channel and the other having a fermionic fusion channel described by the basis (6.57). We now consider the case where we transport one vortex from one pair around another vortex from the other pair. To simplify our calculation we consider loops that do not span any net area in position space. This is illustrated in Figure 6.13(a), where the dashed lines indicate the two oriented parts C_1 and C_2 of the total path $C = C_2^{-1} C_1^{-1} C_2 C_1$. The evolution along this path is cyclic in the space of coupling configurations J_{ij}, where the transport is implemented. The worldline description of the vortices is shown in Figure 6.13(b) that corresponds to exchanging the vortices twice.

The behaviour of Ising anyons undergoing the described evolution is given in Figure 4.12(b). It predicts that the braiding evolution, or monodromy, is given by

$$
B = e^{-\frac{\pi}{4}i} \begin{pmatrix} 0 & 1 \\ 1 & 0 \end{pmatrix}.
\tag{6.58}
$$

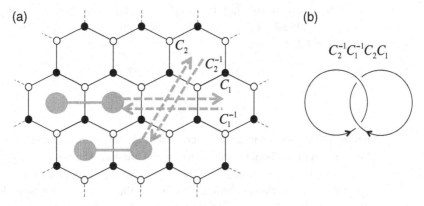

Fig. 6.13 (a) The honeycomb lattice containing two vortex pairs. The four dashed arrows C_1, C_1^{-1}, C_2 and C_2^{-1} are the oriented parts of the path C along which the vortices are transported. (b) $C = C_2^{-1} C_1^{-1} C_2 C_1$ is topologically equivalent to a link.

We shall demonstrate that the geometric evolution that corresponds to the braiding of vortices indeed reproduces this matrix. To numerically evaluate this evolution we need to employ a discrete version of the holonomy presented in Subsection 2.2.2. Let us reiterate the main ingredients. Consider a Hamiltonian $H(\lambda)$ with n-fold degeneracy $\{| \Psi_\alpha(\lambda)\rangle, \alpha = 1, \ldots, n\}$ that depends on some parameters λ. When we vary λ adiabatically along a closed path C, the evolution of the degenerate subspace is given by the holonomy (Wilczek and Zee, 1984)

$$\Gamma_{\mathbf{A}}(C) = \mathbf{P} \exp \oint_C \mathbf{A} \cdot d\lambda, \qquad (A_\mu)_{\alpha\beta}(\lambda) = \langle \Psi_\alpha(\lambda) | \frac{\partial}{\partial \lambda^\mu} | \Psi_\beta(\lambda)\rangle. \qquad (6.59)$$

Let us discretise the path C into T infinitesimal intervals of length $\delta\lambda$, with λ_t denoting the control parameter value at step t. Then we can write

$$\Gamma_{\mathbf{A}}(C) = \lim_{T \to \infty} \mathbf{P} \prod_{t=1}^{T} \left[\mathbb{1} + \delta\lambda^\mu A_\mu(\lambda_t) \right]. \qquad (6.60)$$

Discretising the derivative in $A_\mu(\lambda)$, we have

$$(A_\mu)_{\alpha\beta}(\lambda_t) = \frac{1}{\delta\lambda^\mu} \langle \Psi_\alpha(\lambda_t^\mu) | \Psi_\beta(\lambda_{t+1}^\mu)\rangle - \delta_{\alpha\beta} . \qquad (6.61)$$

Inserting this into the discretised holonomy (6.60), and grouping the states at step t together, we obtain

$$\Gamma_{\mathbf{A}}(C) = \lim_{T \to \infty} \mathbf{P} \prod_{t=1}^{T} \left(\sum_{\alpha=1}^{n} | \Psi_\alpha(\lambda_t)\rangle \langle \Psi_\alpha(\lambda_t)| \right). \qquad (6.62)$$

This is a convenient formula that gives the holonomy as an ordered product of projectors onto the ground state space in the limit $\delta\lambda \to 0$ at each step t along the path C.

One can employ (6.62) to numerically extract the braiding matrix of two vortices by changing the parameters $\{\lambda\} = \{J_{ij}, K_{ijk}\}$ in discrete steps. We now consider the two states, $| \Psi_1\rangle$ and $| \Psi_2\rangle$, corresponding to the single fermion fusion channel (6.57) of the vortices.

The fermionic nature of Hamiltonian (6.51) dictates that these states are given by the Slater determinant (1929) of the vectors $|\psi_i^-\rangle$ denoted in (6.52). In Lahtinen and Pachos (2009) it was found that

$$\Gamma_{\mathbf{A}}(C) \approx B, \qquad\qquad (6.63)$$

within 1% accuracy for suitable parameter regimes. This provides a direct verification that the vortices of the honeycomb lattice model are non-Abelian Ising anyons.

The numerical treatment can investigate the topological nature of the holonomy that describes the braiding evolution. First, when the evolution corresponds to the trivial path $C_0 = C_2 C_2^{-1} C_1 C_1^{-1}$, we obtain $\Gamma_{\mathbf{A}}(C_0) \approx \mathbb{1}$. Second, when the orientation of the braiding is reversed, we obtain inverse evolution, i.e., $\Gamma_{\mathbf{A}}(C^{-1}) = \Gamma_{\mathbf{A}}(C)^{\dagger}$. This follows from the properties of the holonomies described in Section 2.2. Finally, the holonomy is not affected by path deformations $C \to C'$ (i.e., $\Gamma_{\mathbf{A}}(C) = \Gamma_{\mathbf{A}}(C')$) as long as the topology of the path remains the same. All these properties are non-trivial as they only emerge after considering the whole cyclic evolution.

For the evolution of the vortices to correspond to the holonomy $\Gamma_{\mathbf{A}}(C)$, the adiabaticity condition needs to be satisfied at all times and the states $|\Psi_1(\lambda)\rangle$ and $|\Psi_2(\lambda)\rangle$ need to be degenerate. Nevertheless, these states might have a small energy splitting, ε, due to the interactions between vortices. How can we assign $\Gamma_{\mathbf{A}}(C)$ to the evolution of this system? The adiabaticity condition ensures that no population will be transferred between two states, if the evolution is slow enough with respect to their energy difference. What we know is that the energy gap Δ above $|\Psi_1(\lambda)\rangle$ and $|\Psi_2(\lambda)\rangle$ is much larger than their relative energy difference ε. An evolution that is slow enough compared to Δ will adiabatically eliminate the rest of the states. However, if the same evolution is fast enough with respect to ε then it will, in general, transform the Hilbert space of these two states. As long as their relative energy splitting is much smaller than the gap above them, the adiabaticity condition is expected to hold in the transport of vortices and the states $|\Psi_1(\lambda)\rangle$ and $|\Psi_2(\lambda)\rangle$ will effectively behave as degenerate.

Given sufficient site addressability, the presented method of tuning the couplings J can implement vortex transport in a physical realisation of the honeycomb lattice model. The above calculation provides exact predictions for such experiments in finite-size systems. Were the model ever employed for quantum information processing, these studies could be used to predict the fidelities of quantum gates.

6.3 Ising anyons as Majorana fermions

The vortices of the honeycomb lattice model behave as Ising non-Abelian anyons. This hypothesis is supported by the numerical study of the model described in the previous section. Moreover, we found that the continuum limit of this model supports Majorana modes. In particular, zero-energy Majorana modes are expected to be bound on the vortices (Gurarie and Radzihovsky, 2007; Read and Green, 2000). In this section we demonstrate that these Majorana fermions are responsible for the Ising non-Abelian behaviour of the vortices (Ivanov, 2001; Stern *et al.*, 2004; Stone and Chung, 2003). It is worth noticing that the Majorana modes need to be localised in space to interpret the Majorana fermions as Ising anyons. Only then can we attribute particle-like properties to them.

Consider a Majorana operator γ at a given position. Apart from the usual fermionic anticommutation relations, its Majorana character is defined by the 'reality' condition

$$\gamma^\dagger = \gamma. \tag{6.64}$$

In other words, Majorana particles are their own antiparticles. Due to this property it is not possible to define a local degree of freedom for an isolated Majorana mode. However, two such modes localised at positions i and $i + 1$, regardless of how far separated they are, can be combined to a complex fermion mode $z_i = (\gamma_i + i\gamma_{i+1})/2$. The occupation of this mode is a non-local property of a Majorana pair. This scenario emerged in the case of the honeycomb lattice vortices. While the spectrum of that model comprises complex fermionic modes, the presence of pairs of vortices split one such fermion into two localised Majorana modes, like the ones described here.

Let us take four localised Majorana fermions $\gamma_1, \ldots, \gamma_4$, as depicted in Figure 6.14. It is possible to group them in different ways when writing down their Hilbert space. For example, we can consider the two different fermionic representations

$$z_1 = (\gamma_1 + i\gamma_2)/2, \qquad z_2 = (\gamma_3 + i\gamma_4)/2 \tag{6.65}$$

and

$$w_1 = (\gamma_1 + i\gamma_3)/2, \qquad w_2 = (\gamma_2 + i\gamma_4)/2 \tag{6.66}$$

for the same four Majorana fermions. Conversely, we can define the Majorana fermions as 'real' and 'imaginary' decompositions of fermionic modes. A consistent set of fermionic anticommutation relations is given by

$$\{\gamma_i, \gamma_i\} = 2\delta_{ij} \tag{6.67}$$

as well as

$$\{z_i, z_i^\dagger\} = 1 \text{ and } \{w_i, w_i^\dagger\} = 1. \tag{6.68}$$

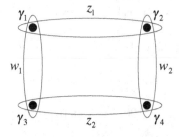

Fig. 6.14 Four localised Majorana fermions $\gamma_1, ..., \gamma_4$. They can be considered as the 'real' and 'imaginary' decompositions of different sets of fermions, e.g., $z_1 = (\gamma_1 + i\gamma_2)/2$, $z_2 = (\gamma_3 + i\gamma_4)/2$ or $w_1 = (\gamma_1 + i\gamma_3)/2$, $w_2 = (\gamma_2 + i\gamma_4)/2$.

It is interesting to note that the modes, z and w, that result from the γ Majorana fermions are not independent. As an example, one can show that

$$\{z_1, w_1\} = \frac{1}{2}. \tag{6.69}$$

We first consider the fusion properties of these Majorana fermions, aiming to derive the corresponding F matrix. We assume that an even number of usual fermionic modes fuse to the vacuum. This is the case, for example, in a superconductor where two fermions are combined to form a Cooper pair. The population states (e.g., of the z fermions) are given by $|ij\rangle_z = |i\rangle_{z_1} \otimes |j\rangle_{z_2}$ with $i, j = 0, 1$, where z_1 has i population and z_2 has j. The operators $z_i^\dagger z_i$ or $w_i^\dagger w_i$ project on the zero population states, while $z_i z_i^\dagger$ or $w_i w_i^\dagger$ project on states that correspond to populated modes. We choose to be initially in the vacuum total fusion space of the two fermionic modes. Then, either both modes z_1 and z_2 are empty or full. The same applies to the w_1 and w_2 modes. The relevant states to the vacuum total fusion channel are given by $|00\rangle_{z/w}$ and $|11\rangle_{z/w}$.

We are now interested to write the states that are initially given in the z basis in terms of states of the w basis. Consider the $|00\rangle_z$ and $|11\rangle_z$ states that satisfy

$$z_1^\dagger z_1 |00\rangle_z = 0, \quad z_1 z_1^\dagger |00\rangle_z = |00\rangle_z, \quad z_1^\dagger z_1 |11\rangle_z = |11\rangle_z, \quad z_1 z_1^\dagger |11\rangle_z = 0. \tag{6.70}$$

As the operators z_1 and w_1 do not anticommute we can ask the question, what is the z population of the state $(2w_1^\dagger w_1 - 1)|00\rangle_z$? Calculating this we find

$$z_1^\dagger z_1 (2w_1^\dagger w_1 - 1)|00\rangle_z = (2w_1^\dagger w_1 - 1)(1 - 2z_1^\dagger z_1)|00\rangle_z = (2w_1^\dagger w_1 - 1)|00\rangle_z, \tag{6.71}$$

where we employed the relation $[z_1^\dagger z_1, w_1^\dagger w_1] = (2w_1^\dagger w_1 - 1)(1 - 2z_1^\dagger z_1)/2$. Hence,

$$|11\rangle_z = (2w_1^\dagger w_1 - 1)|00\rangle_z. \tag{6.72}$$

This calculation stays the same even if we substitute z_2 in place of z_1 and/or w_2 in place of w_1. But the operator $w_1^\dagger w_1$ projects the state $|00\rangle_z$ onto $|11\rangle_w$. Hence, rewriting (6.72) we have

$$|11\rangle_w = \sqrt{2} w_1^\dagger w_1 |00\rangle_z = \frac{1}{\sqrt{2}}(|00\rangle_z + |11\rangle_z). \tag{6.73}$$

The normalisation of the $|11\rangle_w$ state is guaranteed due to the right-hand side of (6.73). We can now employ $\{w_1, w_1^\dagger\} = 1$ to rewrite (6.73) as

$$\sqrt{2} w_1^\dagger w_1 |00\rangle_z = \sqrt{2}(-w_1 w_1^\dagger + 1)|00\rangle_z = \frac{1}{\sqrt{2}}(|00\rangle_z + |11\rangle_z), \tag{6.74}$$

which implies

$$|00\rangle_w = \sqrt{2} w_1 w_1^\dagger |00\rangle_z = -\frac{1}{\sqrt{2}}(|00\rangle_z - |11\rangle_z). \tag{6.75}$$

Thus, from (6.73) and (6.75), we deduce that the w and z bases are related by the unitary transformation

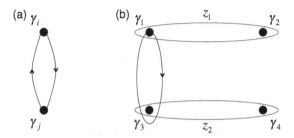

Fig. 6.15 (a) The exchange of two Majorana fermions, γ_i and γ_j. (b) Four Majorana fermions $\gamma_1,..., \gamma_4$ give rise to $z_1 = (\gamma_1 + i\gamma_2)/2$ and $z_2 = (\gamma_3 + i\gamma_4)/2$ fermions. When Majorana fermion γ_1 is braided around Majorana fermion γ_3 then their non-Abelian unitary evolution is given by $\mathcal{U}^2 = e^{2i\theta} \gamma_1\gamma_3 = e^{2i\theta}(z_1z_2 + z_1z_2^\dagger + z_1^\dagger z_2 + z_1^\dagger z_2^\dagger)$.

$$F_M = \frac{1}{\sqrt{2}} \begin{pmatrix} -1 & 1 \\ 1 & 1 \end{pmatrix}. \tag{6.76}$$

This matrix is the fusion matrix of the Ising model up to an exchange of the basis vectors.

Let us now turn to the braiding properties of Majorana fermions. For that consider the clockwise exchange of two Majorana fermions γ_i and γ_j as shown in Figure 6.15(a). The most general unitary operator that can act on their state is given by

$$\mathcal{U} = a\mathbb{1} + b\gamma_i + c\gamma_j + d\gamma_i\gamma_j, \tag{6.77}$$

where a, b, c and d are general complex numbers. The unitarity condition $\mathcal{U}\mathcal{U}^\dagger = \mathbb{1}$ imposes the following equations:

$$aa^* + bb^* + cc^* + dd^* = 1, \quad ab^* + ba^* + cd^* + dc^* = 0,$$
$$ac^* - bd^* + ca^* - db^* = 0, \quad -ad^* + bc^* - cb^* + da^* = 0. \tag{6.78}$$

Moreover, as this evolution is meant to exchange γ_i and γ_j, we should also have $\mathcal{U}\gamma_i\mathcal{U}^\dagger \propto \gamma_j$. This gives

$$aa^* + bb^* - cc^* - dd^* = 0, \quad ab^* + ba^* - cd^* - dc^* = 0,$$
$$-ac^* + bd^* + ca^* - db^* = 0, \quad ad^* - bc^* - cb^* + da^* \neq 0. \tag{6.79}$$

Finally, it should be $\mathcal{U}\gamma_j\mathcal{U}^\dagger \propto \gamma_i$, giving

$$aa^* - bb^* + cc^* - dd^* = 0, \quad -ab^* + ba^* - cd^* + dc^* = 0,$$
$$ac^* + bd^* + ca^* + db^* = 0, \quad ad^* + bc^* + cb^* + da^* \neq 0. \tag{6.80}$$

We can easily verify that there are two independent solutions to these equations. Either $a = d = 0$ and $b = c = e^{i\phi}/\sqrt{2}$ or $b = c = 0$ and $a = d = e^{i\theta}/\sqrt{2}$. These solutions correspond to two distinctive evolutions, $\mathcal{U}$ and $\mathcal{U}'$:

$$\mathcal{U} = \frac{e^{i\theta}}{\sqrt{2}}(\mathbb{1} + \gamma_i\gamma_j), \qquad \mathcal{U}' = \frac{e^{i\phi}}{\sqrt{2}}(\gamma_i + \gamma_j), \tag{6.81}$$

respectively. The second one, $\mathcal{U}'$, is Abelian as $\mathcal{U}'^2 = e^{i2\phi}$. The first one, $\mathcal{U}$, corresponds to a non-Abelian monodromy $\mathcal{U}^2 = e^{2i\theta}\gamma_i\gamma_j$. It is obtained when two Majorana fermions

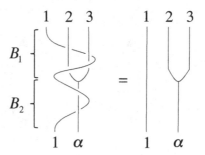

Consider three Majorana fermions, γ_1, γ_2 and γ_3. The monodromy B_1 corresponds to the process where γ_1 braids around γ_2 and γ_3 in a clockwise fashion. Then γ_2 and γ_3 are fused, resulting in the particle α being either the vacuum or a fermion. Finally, γ_1 braids counterclockwise around α, which is described by the monodromy B_2. The resulting evolution is equivalent to the trivial one, i.e., $B_2 B_1 = \mathbb{1}$.

that belong to different fermionic modes are braided. To analyse the action of $\mathcal{U}$ consider the configuration in Figure 6.15(b) with four Majorana fermions γ_i, $i = 1, \ldots, 4$ giving rise to two fermionic modes $z_1 = (\gamma_1 + i\gamma_2)/2$ and $z_2 = (\gamma_3 + i\gamma_4)/2$. Exchanging γ_1 and γ_3 twice gives rise to the monodromy

$$\mathcal{U}^2 = e^{2i\theta}\gamma_1\gamma_3 = e^{2i\theta}(z_1 z_2 + z_1 z_2^\dagger + z_1^\dagger z_2 + z_1^\dagger z_2^\dagger). \tag{6.82}$$

Up to the overall phase factor $e^{2i\theta}$ this operator exchanges the states $|01\rangle_z$ and $|10\rangle_z$ or the states $|00\rangle_z$ and $|11\rangle_z$. Hence, it acts as a σ^x in either of these spaces, which is the expected non-Abelian action of the Ising anyon monodromy.

Finally, we would like to determine the exact value of the phase factor θ. To proceed, we assume that when a Majorana fermion γ and a normal fermion ψ are exchanged clockwise, a phase factor i is acquired. This is in agreement with having two Majorana fermions being the constituents of a usual fermion: when both of them are exchanged with a fermion their total state acquires a factor $i^2 = -1$ resulting from the two i contributions of each exchange of the Majorana with the fermion.

To determine the phase, θ, of the non-Abelian braiding we consider the particular evolution between three Majorana fermions, γ_1, γ_2 and γ_3, as shown in Figure 6.16. Initially γ_1 braids around γ_2 and γ_3 in a clockwise fashion. Then γ_2 and γ_3 are fused and γ_1 braids around the fusion outcome of γ_2 and γ_3 in a counterclockwise fashion.

The composite evolution of these braidings should be equal to the identity as the total braiding procedure is trivial. Indeed, the clockwise braiding between γ_1 and the pair of γ_2 and γ_3 gives the monodromy

$$B_1 = \mathcal{U}_{13}^2 \mathcal{U}_{12}^2 = e^{2i\theta}\gamma_1\gamma_3 e^{2i\theta}\gamma_1\gamma_2 = e^{i4\theta}\gamma_2\gamma_3. \tag{6.83}$$

The counterclockwise braiding of γ_1 with the composite fermion $\alpha = (\gamma_2 + i\gamma_3)/2$ is moreover given by

$$B_2 = 1 - 2\alpha^\dagger\alpha = i\gamma_2\gamma_3. \tag{6.84}$$

This means the braiding evolution B_2 gives 1 if the fermionic mode α is not occupied and -1 if it is occupied. This is in agreement with the fact that the exchange of a Majorana fermion with a normal fermion should give the phase factor i. As deduced from Figure 6.16, the product of the braiding evolutions B_1 and B_2 should be equal to the identity, i.e.,

$$B_2 B_1 = \mathbb{1}. \tag{6.85}$$

Indeed, the worldline of γ_1 in the left-hand side of Figure 6.16 can be deformed continuously to give the right-hand side. Relation (6.85) provides a non-trivial condition for the value of the phase θ, giving

$$\theta = \frac{\pi}{8}. \tag{6.86}$$

This is the same result as one obtains by solving the pentagon and hexagon identities for the Ising model presented in Chapter 4.

Summary

To physically realise a given type of anyon we need to implement the corresponding topological model experimentally. An attractive proposal is Kitaev's honeycomb lattice model presented in this chapter. It employs physically plausible two- and three-spin interactions and gives rise to non-Abelian Ising anyons. This model shares many common features with the $\nu = 5/2$ fractional quantum Hall effect (Moore and Read, 1991), the p-wave superconductors (Chamon *et al.*, 2001) and certain topological insulators (Hasan and Kane, 2010). Hence, studying the honeycomb lattice model offers an arena for addressing generic questions such as the interactions between vortices, the resilience of topological behaviour and so on.

We studied this model by employing two techniques. First, we took the continuum limit of the lattice, which resulted in relativistic Dirac fermions. This approach is similar to the one followed in graphene (Wallace, 1947). When the effective Dirac system has an energy gap, then the Chern number of the model acquires a non-trivial value. This signals that the model can support anyons. As a second approach we employed numerical diagonalisations. This indicated the presence of zero fermionic modes that correspond to Majorana fermions. When studying the spectral evolution as a function of vortex separation, we observed interactions between the zero modes at short distances. This property could be employed for the experimental identification of Majorana fermions by spectral measurements. The transport of one vortex around another revealed their mutual non-Abelian statistics in the form of a geometric phase.

The study of the honeycomb lattice model offers a better understanding of the physics involved with topological models. The microscopic degrees of freedom such as the lattice spins and their interactions provide the knobs to manipulate the emerging anyons and realise their generation, braiding and fusion. This gives a recipe for the experimental implementation of topological quantum computation. While the braiding of Ising anyons alone is not enough to generate a universal set of quantum gates adding interactions between anyons, like the ones we identified here, can result in universality.

Exercises

6.1   Consider a one-dimensional spin chain with interactions $\sigma_i^z \sigma_{i+1}^z$ and $\sigma_i^x \sigma_{i+1}^x + \sigma_i^y \sigma_{i+1}^y$ alternating between pairs of spins, with arbitrary couplings. Diagonalise the Hamiltonian for finite periodic chains and in the thermodynamic limit by employing Majorana fermionisation.

6.2   Consider a honeycomb lattice of tunnelling electrons that models graphene. What are the similarities and what are the differences of the continuum limit of the graphene system compared to the continuum limit of the honeycomb lattice model of Majorana fermions?

6.3   Derive the general form of the interactions (6.51).

Chern–Simons quantum field theories

Quantum field theory is the most efficient tool we have to describe elementary particles. It is the backbone of the Standard Model that successfully explains electromagnetic, weak and strong interactions. For example, quantum electrodynamics, which describes the interactions between charged fermions mediated by photons within the field theory framework, has been tested experimentally and verified to a high level of accuracy. This has established quantum field theory as the definitive tool for studying high-energy physics.

Apart from explaining the fundamental properties of matter, quantum field theories can also provide an effective description of condensed matter systems. In Chapter 6 we saw that the quantum field theory of Majorana fermions emerges from Kitaev's honeycomb lattice model. It is expected that the low-energy behaviour of the fractional quantum Hall effect can be described efficiently by certain quantum field theories, known as Chern–Simons theories (Froehlich *et al.*, 1997). The fractional quantum Hall effect emerges in interacting two-dimensional electron gases at low temperature in the presence of a strong magnetic field. Due to its complexity, this system evades exact theoretical analysis. An effective description with Chern–Simons theories has nevertheless proven very fruitful in understanding its topological properties.

Chern–Simons theories are topological quantum field theories in the sense that all their observables are invariant under continuous coordinate transformations. In other words, relative distances, and subsequently local geometry, do not play a role in these theories. This makes the resulting interactions rather special: meaningful quantities are functions only of global topological characteristics. Unlike usual electromagnetic interactions their effect does not depend on the relative position of charges. It is rather similar in nature to the Aharonov–Bohm effect that emerges when topologically non-trivial evolutions take place. This is exactly the framework that gives rise to anyonic particles. Indeed, Chern–Simons theories provide yet another model that supports Abelian as well as non-Abelian anyons with their own fusion rules and braiding properties.

The relation between Chern–Simons theories and the fractional quantum Hall effect is the subject of current theoretical and experimental research. Here we do not aim to analyse this relation, but we present the Chern–Simons theories and their anyonic content in detail. Studying these theories opens the way to the next chapter that deals with the efficient evaluation of topological invariants of links with quantum algorithms.

In the following we initially present the Abelian Chern–Simons theories. Using classical and quantum mechanical arguments we demonstrate that their sources behave as Abelian anyons. Subsequently, non-Abelian Chern–Simons theories are analysed. The fusion and braiding properties of their sources are presented and it is shown that they correspond to a discrete, infinite family of non-Abelian anyonic models.

7.1 Abelian Chern–Simons theories

The Abelian Chern–Simons theories arise as a possible generalisation of electromagnetic theory when spacetime is reduced from four to three dimensions. It gives a simple formalism, where particles with Abelian anyonic statistics emerge. These anyons are manifested in the form of sources that interact through the Chern–Simons gauge field. Intriguingly, quantum amplitudes that describe anyonic evolutions are actually given in terms of simple topological invariants of links.

7.1.1 Four-dimensional electromagnetism

It is instructive to start from the usual electromagnetism in three spatial and one time dimensions. In our analysis we employ the action, which is equivalent to the Hamiltonian formalism. In particular, we are concerned with the behaviour of the theory in the presence of external sources. From the stationary property of the action we can derive the classical equations of motion, i.e., Maxwell equations. Alternatively, from the path integral formulation we can obtain quantum amplitudes that describe evolutions of external sources.

The action of the four-dimensional electromagnetic field coupled to currents and charges is given by

$$I_{\mathrm{EM}}^{\mathrm{4d}}[A] = \frac{1}{2} \int d^4x \left(\mathbf{E}^2 - \mathbf{B}^2 + \mathbf{A} \cdot \mathbf{J} + A_0 \rho \right). \tag{7.1}$$

Here $\mathbf{E}$ and $\mathbf{B}$ are the electric and magnetic fields, respectively, $\mathbf{J}$ is the charge current, ρ is the charge density and $x = (t, \mathbf{x}) = (x^0, x^1, x^2, x^3)$ parameterises spacetime. The fields couple to the current and the charge via the corresponding vector potential $\mathbf{A}$ that satisfies $\mathbf{B} = \nabla \times \mathbf{A}$ and the scalar potential A_0 that satisfies $\mathbf{E} = -\nabla A_0$. This homogeneous and isotropic action is the most general one that is Lorentz- and gauge-invariant (Jackson, 1975). The resulting equations of motion are given by

$$\nabla \cdot \mathbf{E} = \rho, \quad \nabla \times \mathbf{B} - \frac{\partial \mathbf{E}}{\partial t} = \mathbf{J}. \tag{7.2}$$

Together with the identities

$$\nabla \cdot \mathbf{B} = 0, \quad \nabla \times \mathbf{E} + \frac{\partial \mathbf{B}}{\partial t} = 0, \tag{7.3}$$

they constitute Maxwell equations. Maxwell equations describe the interactions between charges and currents mediated by electromagnetic fields. In particular, they give rise to the long-range Coulomb interaction between two charges at distance r that scales as $1/r^2$. This long-range behaviour of the electromagnetic interactions is a result of the massless nature of the gauge field $(A_0, \mathbf{A})$. The absence of any mass term of the gauge field in the action (7.1) is imposed by the invariance under gauge transformations of the form $A_0 \to A_0^\omega = A_0 + \partial_t \omega$ and $\mathbf{A} \to \mathbf{A}^\omega = \mathbf{A} + \nabla \omega$, where ω is an arbitrary scalar function. We have seen in Subsection 2.1.2 that electromagnetism can give rise to topological interactions. For that we need to arrange charges and fluxes, so that the Aharonov–Bohm effect emerges. In practice, it is hard to mechanically isolate the topological from the direct interactions.

7.1.2 Three-dimensional electromagnetism

Reducing the spatial dimensions from three to two actually allows the electromagnetic theory to have a richer structure. Let us look for the most general form of Maxwell equations in three dimensions, with coordinates $x = (t, \mathbf{x}) = (x^0, x^1, x^2)$, where space is considered as a flat plane. In this case the electric field has only two components, $\mathbf{E} = (E_1, E_2)$, which are parallel to the two-dimensional plane. The magnetic field is only a scalar, B, that corresponds to the third component of the magnetic field if it were thought to emerge from three spatial dimensions. Still the vector potential is a vector $\mathbf{A} = (A_1, A_2)$. In two spatial dimensions the curl of a vector $\mathbf{V} = (V_1, V_2)$ becomes a scalar,

$$\mathbf{\nabla} \times \mathbf{V} = \frac{\partial V_2}{\partial x^1} - \frac{\partial V_1}{\partial x^2}, \tag{7.4}$$

and the curl of a scalar S becomes a vector,

$$\mathbf{\nabla} \times S = \left(\frac{\partial S}{\partial x^2}, -\frac{\partial S}{\partial x^1} \right). \tag{7.5}$$

The most general Lorentz- and gauge-invariant action in three spacetime dimensions that is also homogeneous and isotropic is given by

$$I_{\text{EM}}^{3d}[A] = \frac{1}{2} \int d^3x \left(\mathbf{E}^2 - B^2 + \mathbf{A} \cdot \mathbf{J} + A_0 \rho + m\varepsilon^{\mu\nu\rho} A_\mu \partial_\nu A_\rho \right). \tag{7.6}$$

We notice now that an additional m-term is allowed. This is the Chern–Simons action given by

$$I_{\text{CS}} = \frac{m}{2} \int d^3x \varepsilon^{\mu\nu\rho} A_\mu \partial_\nu A_\rho. \tag{7.7}$$

Due to the reduced dimensionality this action is invariant under gauge transformations of the vector potential $\mathbf{A}$ (see Exercise 7.1). Moreover, it does not depend on a metric that would have introduced a measure of distance in the Chern–Simons theory. Such a metric $g_{\mu\nu}$ is present in (7.6) as, for example, it is used in $\mathbf{E}^2 = g_{\mu\nu} E^\mu E^\nu$. Instead, the three-dimensional Levi–Civita symbol $\varepsilon^{\mu\nu\rho}$ is employed in (7.7), whose indices run through 0, 1 and 2. The Levi–Civita symbol is equal to zero if two or more indices are equal to each other, 1 if the indices are any cyclic permutation of $(0, 1, 2)$ and -1 otherwise. This object and consequently the Chern–Simons action is manifestly invariant under coordinate transformations. The independence of the action on relative distances is a defining characteristic of a topological quantum field theory.

The additional m-term in (7.6) gives rise to massive 'photons', where m is a constant with the dimensions of a mass. Indeed, the m-term makes the theory similar to a quantum field theory of massive fields (Weinberg, 1995). Unlike the case of massless photons the fields in the massive case decay exponentially fast, e^{-mr}. This short-range behaviour is also encountered in the weak interactions of the Standard Model which are mediated by the Z or W massive bosons. The exponential damping of the interactions restricts both fields $\mathbf{E}$ and B to take non-trivial values only at the immediate neighbourhood of the sources. It is a particular characteristic of the mass term (7.7) that, due to the presence of the derivative, the field $\mathbf{A}$ can still take large values away from the sources, much like in the Aharonov–Bohm effect.

7.1.3 Abelian anyons and topological invariants

To present how Abelian anyons emerge in the Abelian Chern–Simons theories we employ two approaches. First, with simple classical arguments we show how charged sources are necessarily accompanied by flux. This charge–flux composition is responsible for the anyonic behaviour of the sources. Then we employ a quantum approach that gives equivalent results through a very different line of reasoning. Its purpose is to establish the connection between the Chern–Simon theories and topological invariants of links. Later on we use the same quantum approach to study non-Abelian Chern–Simons theories.

7.1.3.1 Classical approach

To classically analyse the model we derive the equations of motion corresponding to the action (7.6). The resulting three-dimensional Maxwell equations are given by

$$\nabla \times \mathbf{E} + \frac{\partial B}{\partial t} = 0, \tag{7.8}$$

$$\nabla \cdot \mathbf{E} + mB = \rho, \tag{7.9}$$

$$\nabla \times B - \frac{\partial \mathbf{E}}{\partial t} + m \begin{pmatrix} E^2 \\ -E^1 \end{pmatrix} = \mathbf{J}. \tag{7.10}$$

In particular, let us consider (7.9). It dictates that apart from the usual electric field a magnetic field is generated from the static charge density ρ. Both the electric and magnetic fields are exponentially localised near charges. In fact, if we neglect the electric field term, we see that the magnetic field exactly follows the distribution of the charge density, ρ. By integrating (7.9) over the whole space we have

$$\int d^2\mathbf{x}\nabla \cdot \mathbf{E} + m \int d^2\mathbf{x}B = \int d^2\mathbf{x}\rho. \tag{7.11}$$

Due to Stokes' theorem the first integral becomes a line integral of the electric field at the boundary. As the field $\mathbf{E}$ does not extend to large distances away from the sources, this integral gives zero. The second integral is the total flux Φ that is confined at the position of the source. The right-hand side is the total charge Q. As a conclusion, for $m \neq 0$ we have

$$\Phi = \frac{Q}{m}, \tag{7.12}$$

which shows that every charged particle carries a magnetic flux. When $m = 0$, the mechanism of attaching flux to charged particles does not exist.

The emerging flux (7.12), which accompanies the charge Q, can be considered as a background magnetic field that is allowed to take a non-zero value at the position of the charged source. In other words, the flux Φ does not correspond to magnetic dipoles that get transported with the charge. This behaviour is different from the flux and charge of composite anyons we met in Subsection 2.1.3. There, the flux solenoids were created from magnetic dipoles that could acquire phases from the Aharonov–Bohm effect if they were transported around charges.

From (7.12) we see that the Chern–Simons term enables a specific flux to be assigned to every charged particle. At the same time it neutralises direct, position-dependent interactions between charges by making them short range. These properties define the statistics of sources in $I_{\mathrm{EM}}^{\mathrm{3d}}$. Consider two identical sources 1 and 2, which are described by disks with homogeneously distributed charge Q. We want to determine the quantum evolution resulting from braiding source 1 around source 2. If these sources are kept far apart the only contribution to the quantum evolution of their state is topological in nature, due to the Aharonov–Bohm effect. A circulation, which corresponds to two braidings of such particles, hence gives the phase factor

$$2\varphi = Q\Phi = \frac{Q^2}{m}, \tag{7.13}$$

where φ is the statistical angle. Due to the spin-statistics theorem each source carries spin (see Exercise 7.2)

$$s = \frac{Q^2}{4\pi m}. \tag{7.14}$$

From (7.13) and (7.14) we see that the Chern–Simons theories can support anyons with arbitrary abelian statistics as the parameter m can take any value.

7.1.3.2 Quantum approach

The anyonic properties of charges can also be revealed with a quantum treatment of the Abelian Chern–Simons theories. This approach illuminates the role topological invariants play in Chern–Simons theories. To show that, we adopt the Euclidean spacetime region M, shown in Figure 7.1. This region is taken in general to have trivial topology with boundary ∂M positioned far from the worldlines of the particles. Suppose we want to develop the quantum field theory of a U(1) gauge field with components $A = (A_0, A_1, A_2)$. To do so we focus on the Abelian Chern–Simons action

$$I_{\mathrm{CS}}[A] = \frac{m}{2} \int_M d^3x \varepsilon^{\mu\nu\rho} A_\mu \partial_\nu A_\rho, \tag{7.15}$$

which is invariant under continuous coordinate reparameterisations and gauge transformations.

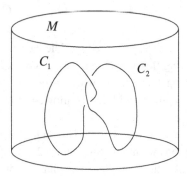

Fig. 7.1 The three-dimensional Euclidean space M, where particle worldlines form a link. Two loops C_1 and C_2 are shown for which Wilson lines can be defined. This diagram can be interpreted as four particle sources that are pairwise created from the vacuum, one source from one pair is braided around another source from the other pair and then the sources are fused back to the vacuum.

To obtain information about the system we need to identify the appropriate observable operators and evaluate their expectation value. To respect the symmetries of the Chern–Simons theories, these operators need to be gauge-invariant and they should not depend on a metric. The product of Wilson loops,

$$W(L) = \prod_{i=1}^{r} \exp\left(iQ_i \oint_{C_i} A \right), \tag{7.16}$$

serves this purpose well. It corresponds to the looping trajectories C_i of particles parameterised by i with charge Q_i, where $i = 1, \ldots, r$. These trajectories, defined inside the region M, are collectively denoted L. They can be interlinked, as the example in Figure 7.1 illustrates. Note that requiring the theory to be invariant under global gauge transformations of the gauge field demands that the charges Q_i are integer numbers (Yang, 1970).

In the path integral formulation the expectation value of the Wilson loops, $W(L)$, is given by

$$\langle W(L) \rangle = \frac{\langle \psi_0 \mid W(L) \mid \psi_0 \rangle}{\langle \psi_0 | \psi_0 \rangle} = \frac{\int \mathcal{D}A W(L) e^{iI_{\text{CS}}[A]}}{\int \mathcal{D}A e^{iI_{\text{CS}}[A]}}, \tag{7.17}$$

where $| \psi_0 \rangle$ is the vacuum state. The functional integral $\int \mathcal{D}A$ is taken with respect to configurations of the gauge field, A, throughout M that are gauge-inequivalent. This expectation value can be interpreted as the amplitude of the process associated with the link L, such as the one depicted in Figure 7.1. There, particles are created in pairs from the vacuum, propagated along the worldlines C_1 and C_2 and then fused to the vacuum so that the worldlines correspond to the link configuration L.

To evaluate the expectation value $\langle W(L) \rangle$ we interpret the Wilson loops in terms of currents. Consider a single loop C spanned by a particle with charge Q. We parameterise the loop by s with $0 \le s \le 1$, so that $y^\mu(s)$ runs once through the loop C. Then we can write

$$Q \oint_C dx^\mu A_\mu(x) = \int_M d^3x J^\mu(x) A_\mu(x), \tag{7.18}$$

with

$$J^\mu(x) = Q \int_0^1 ds\, \dot{y}^\mu(s)\delta(x - y(s)). \tag{7.19}$$

The delta function restricts the integration argument x to be on the points $y(s)$ of C. The integration along s together with $\dot{y}^\mu(s)$ becomes the line integral present in (7.18). To evaluate the functional integral (7.17) we rewrite the gauge field as

$$A_\mu(x) = A'_\mu(x) + \tilde{A}_\mu(x), \tag{7.20}$$

where $A'_\mu(x)$ can take any arbitrary field configuration, while $\tilde{A}_\mu(x)$ is a specific classical configuration. So, in the functional integral, we have $\mathcal{D}A = \mathcal{D}A'$. The classical field configuration is taken to satisfy the Maxwell equations (7.9) and (7.10), i.e.,

$$m\varepsilon^{\mu\nu\rho}\partial_\nu\tilde{A}_\rho(x) = -J^\mu(x), \tag{7.21}$$

with $J^0 = -\rho$ and $(J^1, J^2) = \mathbf{J}$. This condition helps to decouple the contribution of the sources and thus simplifies the calculation. Substituting (7.20) in (7.17) we have

$$\frac{\int \mathcal{D}A\, W(L)e^{il_{\mathrm{CS}}[A]}}{\int \mathcal{D}A\, e^{il_{\mathrm{CS}}[A]}} = \exp\left(\frac{i}{2}\int_M d^3x\, J^\mu(x)\tilde{A}_\mu(x)\right), \tag{7.22}$$

where the functional integrals in the numerator and denominator cancel out. Hence, to evaluate $\langle W(L)\rangle$ we need to determine $\tilde{A}_\mu(x)$. A solution of equation (7.21) is given by (Polyakov, 1988)

$$\tilde{A}_\mu(x) = -\frac{Q}{4\pi m}\int_0^1 ds\, \varepsilon^{\mu\nu\rho}\dot{y}^\nu(s)\frac{(x - y(s))^\rho}{|x - y(s)|^3}, \tag{7.23}$$

where y^μ are the Euclidean coordinates of the loops in M.

The product of Wilson loops, $W(L)$, gives rise to several currents, one for each path C_i. In the presence of many currents the total field configuration $\tilde{A}$ that satisfies (7.21) is given by the sum of the corresponding fields produced from each current. This implies that

$$\int_M d^3x\, J^\mu(x)\tilde{A}_\mu(x) = -\frac{1}{4\pi m}\sum_{i,j} Q_iQ_j \oint_{C_i} dx^\mu \oint_{C_j} dy^\nu \varepsilon^{\mu\nu\rho}\frac{(x - y)^\rho}{|x - y|^3}. \tag{7.24}$$

As a conclusion, the expectation value of the Wilson loops is given by

$$\langle W(L)\rangle = \exp\left(\frac{i}{2m}\sum_{i,j} Q_iQ_j\Phi(C_i, C_j)\right), \tag{7.25}$$

where $\Phi(C_i, C_j)$ is the Gauss integral given by

$$\Phi(C_i, C_j) = \frac{1}{4\pi}\oint_{C_i} dx^\mu \oint_{C_j} dy^\nu \varepsilon^{\mu\nu\rho}\frac{(x - y)^\rho}{|x - y|^3}. \tag{7.26}$$

Hence, we managed to write the expectation value of the Wilson loops in terms of a familiar quantity, the Gauss integral. This integral is a well-defined integer when the loops C_i and C_j do not intersect (Polyakov, 1988). It is the topological invariant of the linking between the two loops, i.e., it counts the number of times C_i winds C_j. For example, the two links in Figure 7.1 have

$$\Phi(C_1, C_2) = \pm 1, \tag{7.27}$$

where the sign depends on the relative orientation of the loops. The topological character of the Gauss linking number, $\Phi(C_i, C_j)$, results from the independence of the action and of the Wilson loops on the metric.

As the Wilson loops correspond to currents of charges, equation (7.25) can be interpreted as describing the behaviour of sources. For example, entangled loops correspond to braided sources. Hence, the closed form of $\langle W(L) \rangle$ given in (7.25) can reveal the anyonic character of the sources. Following Figure 7.1 we see that the quantity $\langle W(L) \rangle$ is the expectation value of four sources being pairwise generated from the vacuum, then one source from one pair is braided with the source from the other pair, and then they are fused back to the vacuum. Let us denote the trivial link that corresponds to two disentangled loops as L_0. Then, according to (7.27), the braiding operation gives the phase factor

$$\langle W(L) \rangle = \exp\left(i\frac{Q^2}{m}\right) \langle W(L_0) \rangle, \tag{7.28}$$

where the summation in (7.25) gave two contributions. Hence, the sources have anyonic mutual statistics, which is in agreement with (7.13).

Close inspection shows that (7.25) is not well defined when $i = j$, as the points x and y in the denominator can coincide. To resolve this problem we adopt the framing prescription (Witten, 1989). According to this prescription we take every link component, C_i, and slightly displace it along a given direction, thus producing a new link component C_i', as shown in Figure 7.2(a). The methodology we adopt is to substitute each term with $i = j$ with a term that has two such distinguishable loops, C_i and C_i', for which the Gauss linking number is well defined. The resulting structure is a ribbon made out of the two loops C_i and C_i'.

The framing procedure not only resolves the mathematical ambiguity of (7.25), but also introduces spin degrees of freedom that make the Chern–Simons theories self-consistent. Indeed, the ribbons provide the means to identify possible twists which cannot be registered by a string. To demonstrate that, consider the expectation value $\langle W(L) \rangle$. When a single ribbon with boundaries C and C' is twisted by 2π, then the linking number between the

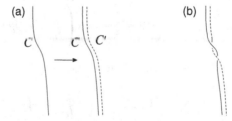

Fig. 7.2   (a) The framing of a string is defined by infinitesimally displacing the original string, C, to create a new string, C'. Both strings together define the edges of a ribbon. (b) A twist is well defined with a ribbon and it gives a non-trivial linking between the two strings, C and C'.

distinguishable links C and C' is changed by one, as shown in Figure 7.2(b). From (7.25) a phase change in the Wilson loop expectation value is recorded such that

$$\langle W(C') \rangle = \exp\left(i\frac{Q^2}{2m}\right) \langle W(C) \rangle. \tag{7.29}$$

This corresponds to having particles with spin

$$s = \frac{Q^2}{4\pi m}, \tag{7.30}$$

in agreement with (7.14).

The above analysis shows that the braiding between sources can be described efficiently by the expectation value of Wilson loops. This expectation value corresponds to the amplitude of anyonic sources which are pairwise created from the vacuum, evolved along world-lines that correspond to the loops of the Wilson operators and afterwards fused back to the vacuum. The evaluation of this amplitude demonstrates that the sources of the Abelian Chern–Simons theories have anyonic statistics.

7.2 Non-Abelian Chern—Simons theories

Our study of non-Abelian Chern–Simons theories follows the approach we took in the Abelian case in terms of expectation values of Wilson loop operators. Unlike the Abelian case, gauge-invariance of the non-Abelian theories is not identically satisfied. To restore gauge-invariance the coupling of the theory needs to be quantised. We then concentrate on the anyonic properties of source particles, identify their fusion rules and illustrate their non-Abelian statistics (Guadagnini *et al.*, 1990; Witten, 1989).

7.2.1 Non-Abelian gauge theories

As a first step we define the action of the non-Abelian Chern–Simons theories and investigate some of its basic properties. The central result of this subsection is that demanding gauge-invariance of the action restricts the theories to a discrete, but infinite set.

7.2.1.1 Non-Abelian Chern–Simons action

The non-Abelian version of the Chern–Simons theories is based on a Lie group G with elements that do not all commute with each other. That is, for two elements $g, h \in G$ in general $gh \neq hg$. Consider the generators of this group given by the Hermitian matrices T^a for $a = 1, \ldots, n$. A general element, $g \in G$, is given by $g = \exp(i\lambda^a T^a)$ for some real parameters λ^a. The generators satisfy the Lie algebra commutation relations

$$[T^a, T^b] = if^{abc}T^c, \tag{7.31}$$

where a summation is assumed over all n values of the repeated index c. The parameters $\{f^{abc}\}$ are the structure constants of the Lie algebra. These matrices satisfy the trace condition

$$\text{tr}(T^a T^b) = \frac{1}{2}\delta^{ab}. \tag{7.32}$$

There are many matrix representations of the Lie algebra that can satisfy (7.31) and (7.32) of different dimension. Later on we consider these representations in more detail. The vector potential of the theory is defined by

$$A_\mu(x) = A_\mu^a(x)T^a. \tag{7.33}$$

Hence, A is a vector defined in the three-dimensional Euclidean space M with components that are Hermitian matrices.

The action of the gauge field is given by

$$I_{\text{CS}}[A] - \frac{k}{4\pi}\int_M d^3x\,\varepsilon^{\mu\nu\rho}\,\text{tr}\Big(A_\mu\partial_\nu A_\rho + i\frac{2}{3}A_\mu A_\nu A_\rho\Big), \tag{7.34}$$

which can also be written as

$$I_{\text{CS}}[A] = \frac{k}{8\pi}\int_M d^3x\,\varepsilon^{\mu\nu\rho}\,\Big(A_\mu^a\partial_\nu A_\rho^a - \frac{1}{3}f^{abc}A_\mu^a A_\nu^b A_\rho^c\Big). \tag{7.35}$$

This expression coincides with the Abelian case when $f^{abc} = 0$, when the field A_μ^a has only a single algebra component (i.e., $n = 1$) and when $k = 4\pi m$.

7.2.1.2 Gauge-invariance

Compared to their Abelian counterparts, non-Abelian gauge fields allow for more complex gauge transformations:

$$A_\mu \to A_\mu^U = U^\dagger A_\mu U - iU^\dagger\partial_\mu U, \tag{7.36}$$

where $U(x) \in G$. Apart from the addition of a term, $U^\dagger\partial_\mu U$, the vector potential obtains a conjugation with the group element U that is purely a result of the non-Abelian character of the group G. Under this gauge transformation the action $I_{\text{CS}}[A]$ becomes

$$I_{\text{CS}}[A^U] = I_{\text{CS}}[A] + i\frac{k}{4\pi}\int_M d^3x\varepsilon^{\mu\nu\rho}\partial_\mu\text{tr}\big(\partial_\nu UU^\dagger A_\rho\big)$$

$$+ \frac{k}{12\pi}\int_M d^3x\varepsilon^{\mu\nu\rho}\text{tr}\big(U^\dagger\partial_\mu UU^\dagger\partial_\nu UU^\dagger\partial_\rho U\big). \tag{7.37}$$

The second term on the right-hand side is an integral of a total derivative. Hence, it becomes a surface integral which vanishes when A is zero at the boundary of M. The last term is proportional to the winding number $\omega[U]$ of the gauge transformation $U(x)$, which is given by

$$\omega[U] = \frac{1}{24\pi^2}\int_M d^3x\varepsilon^{\mu\nu\rho}\text{tr}\big(U^\dagger\partial_\mu UU^\dagger\partial_\nu UU^\dagger\partial_\rho U\big). \tag{7.38}$$

The winding number $\omega[U]$ is an integer that counts how many times the gauge transformation U spans the whole group, G (see also Example II later in this chapter). It is topological in nature and invariant under smooth deformations of $U(x)$.

Let us investigate some of the properties of the winding number, $\omega[U]$. Consider the gauge transformation $U(x)$ as a function of the three-dimensional coordinate x. For $U(x)$ to be a permissible gauge transformation it has to have a constant value when x is taken to infinity. Hence, we choose

$$\lim_{|x| \to \infty} U(x) = \mathbb{1}, \tag{7.39}$$

without loss of generality. As the gauge group element $U(x)$ is defined throughout the whole three-dimensional space $\mathbb{R}^3$ including infinity, its argument space is equivalent to the three-dimensional sphere S^3. It hence provides a mapping from S^3 to the parametric space of the group G. For non-Abelian compact gauge groups such a mapping can be labelled by an integer number that counts how many times the mapping winds around the group G when the whole S^3 is spanned. For a given $U(x)$ we thus have that

$$\omega[U] = n, \tag{7.40}$$

where n is the winding integer. This number is characteristic of the group element $U(x)$ and cannot be changed by local continuous coordinate transformations. This is expected, as the definition of $\omega[U]$ does not involve the metric.

As a result of the above analysis we see that the non-Abelian Chern–Simons action (7.34) acquires an additional term when the field $A_\mu(x)$ is gauge transformed (Deser *et al.*, 1982). In particular, a gauge transformation $U(x)$ with winding number $\omega[U] = n$ gives

$$I_{\text{CS}}[A^U] = I_{\text{CS}}[A] + 2\pi k n. \tag{7.41}$$

A gauge theory is deemed physical, if the expectation values of all of its observables are gauge-invariant. In view of (7.17) we require that $\exp(iI_{\text{CS}}[A])$ is gauge-invariant under a gauge transformation. The necessary condition for that is

$$k : \text{ integer.} \tag{7.42}$$

This means, every non-Abelian Chern–Simons theory is parameterised by a coupling k that only takes integer values. The coupling k is also known as the level of the theory. A quantisation condition like the one in (7.42) was not present in the Abelian case.

7.2.2 Wilson loops and anyonic worldlines

We now want to evaluate the expectation value of a non-Abelian Wilson loop operator with an arbitrary link configuration. For concreteness we consider the group SU(N). The trace of a Wilson loop is given by

$$W(C) = \text{tr}\left[\mathbf{P} \exp\left(iQ \oint_C T^a A_\mu^a dx^\mu \right) \right], \tag{7.43}$$

where the trace is taken with respect to the representation of the gauge field A_μ and $\mathbf{P}$ is the path-ordering symbol. The coupling constant $Q = 2j$ is an integer (see Example III later in this chapter). For convenience, we define $T^a_{(j)} = 2jT^a$ where T^a is an element of the $(2j + 1)$-dimensional representation of the su(N) algebra that we denote R_j. The N-dimensional representation of su(N) is called fundamental.

The trace of a Wilson loop is gauge-invariant. Indeed, under the gauge transformation (7.41) we have

$$\mathbf{P}\exp\left(i\oint_C T^a_{(j)}A^a_\mu{}^U dx^\mu\right) = U(x_0)^\dagger \mathbf{P}\exp\left(i\oint_C T^a_{(j)}A^a_\mu dx^\mu\right) U(x_0), \qquad (7.44)$$

where x_0 is the base point where the path C begins and ends. Taking the trace of (7.44) cancels the $U(x_0)$ dependence, thus making $W(C)$ invariant under gauge transformations. Consider the expectation value of a product of Wilson loops,

$$\langle W(L)\rangle = \frac{\int \mathcal{D}A \prod_{i=1}^r \mathrm{tr}\left[\mathbf{P}\exp\left(i\oint_{C_i} T^a_{(j)}A^a_\mu dx^\mu\right)\right] e^{il_{CS}[A]}}{\int \mathcal{D}A e^{il_{CS}[A]}}. \qquad (7.45)$$

The A integration is with respect to all fields that are not gauge-equivalent to each other as the integrand is manifestly gauge-invariant. Moreover, we restrict the integration to the gauge field configurations that satisfy the classical version of the Gauss law. Hence, A corresponds to a non-Abelian magnetic field $B^a = \partial_1 A^a_2 - \partial_2 A^a_1 - f^{abc}A^b_1 A^c_2$, which is zero everywhere apart from the position of the loops C_i, where sources in the form of the representations R are present (Witten, 1989).

Much like in the Abelian case, we can view the expectation value of Wilson loops as the expectation value of the evolution of non-Abelian anyons. In this picture, the worldlines of the anyons correspond to the paths C_i of the Wilson loops. The type of anyon that traverses each loop C_i corresponds to the representation R_i of the Wilson loop $W(C_i)$. To establish this interpretation we choose the space M to be a spatial disc Σ that extends from $t = -\infty$ to $t = +\infty$, as shown in Figure 7.3. Relation (7.45) then gives the quantum mechanical expectation value of the process where particles are created from the vacuum state $|\psi_0\rangle$ on

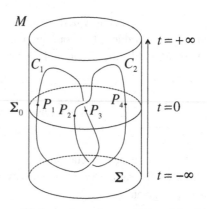

Fig. 7.3 Consider the three-dimensional space M being made up of a spatial disk Σ defined though the infinite time line. A disc Σ_0 at $t = 0$ introduces a set of points P_1, P_2, P_3 and P_4 where the link made out of the loops C_1 and C_2 intersects the disk.

the disk Σ, then they are evolved so that their worldlines traverse the link L and fuse to the vacuum channel. In other words we have

$$\langle W(L) \rangle = \langle \psi_0 | \mathcal{U}(t = -\infty, t = +\infty) | \psi_0 \rangle, \qquad (7.46)$$

where the unitary $\mathcal{U}(t = -\infty, t = +\infty)$ corresponds to the evolution operator of the anyons.

A slice Σ_0 of space M at $t = 0$ gives a particular configuration of sources, as shown in Figure 7.3. Assume that Σ_0 cuts the loops at the points P_i, with each of them corresponding to a Wilson loop in representation R_i. The points P_i can be interpreted as the positions of the sources at time $t = 0$ and the representations R_i can be interpreted as their charge. In the following we determine the fusion and braiding properties of these sources, thereby demonstrating that they behave as non-Abelian anyons.

7.2.2.1 Sources, representations and fusion rules

We now determine the fusion properties of the sources of the non-Abelian Chern–Simons theories. For that we first introduce some basic elements of representation theory (Fulton and Harris, 1991). We then present the composition relation of representations. Finally, we establish how these relations apply to the case of the Chern–Simons theory. In particular, we show that the number of distinct representations in a given theory is fixed by the Chern–Simons coupling k. This fixes the number of possible anyonic species supported by the theory as well as their fusion rules.

A representation of a group G assigns to every element $g \in G$ a square complex matrix $R(g)$ of a given dimension that preserves the group multiplication structure

$$R(g_1 g_2) = R(g_1) R(g_2). \qquad (7.47)$$

Each matrix describes a linear transformation acting on a vector space. The elements of this space can be viewed as column vectors of the appropriate dimension. A matrix representation and its corresponding vector space have equivalent properties related to their tensor product and its decomposition.

As a concrete example we take the SU(2) group. Physically, this group can describe angular momentum or spin transformations. The generators of SU(2) satisfy the algebra

$$[J_a, J_b] = i\epsilon^{abc} J_c. \qquad (7.48)$$

As the total angular momentum operator $J^2 = J_1^2 + J_2^2 + J_3^2$ commutes with J_3, we can denote their common eigenstates as $|j, m_j\rangle$. These satisfy

$$J^2 |j, m_j\rangle = j(j+1) |j, m_j\rangle \quad \text{and} \quad J_3 |j, m_j\rangle = m_j |j, m_j\rangle. \qquad (7.49)$$

Here j is a non-negative integer or half-integer and the corresponding m_j is an integer or half-integer, respectively, that takes values between $-j \leq m_j \leq j$ in integer steps. An SU(2) Chern–Simons source described by a representation with a given j has an associated $(2j+1)$-dimensional Hilbert space generated by the orthonormal basis states $|j, m_j\rangle$, which are parameterised by m_j. The $j = 0$ with $m_j = 0$ case is assigned to the vacuum, for which the corresponding representation is the trivial one-dimensional matrix.

To describe the presence of two sources, 1 and 2, we need to employ the tensor product structure. Indeed, each source is allocated an independent Hilbert space, with spin-like states $\left|j_1, m_{j_1}\right\rangle$ and $\left|j_2, m_{j_2}\right\rangle$. A basis of their composite Hilbert space can be given as the tensor product $\left|j_1, m_{j_1}\right\rangle \otimes \left|j_2, m_{j_2}\right\rangle$. We would like to write the tensor product of these states as another spin state, $\left|j, m_j\right\rangle$. Vector composition of angular momenta dictates that j can take all possible values with $|j_1 - j_2| \leq j \leq j_1 + j_2$ in integer steps. To account for all these possibilities we use the notation

$$j_1 \times j_2 = \sum_{j=|j_1-j_2|}^{j_1+j_2} j. \tag{7.50}$$

An example is the tensor product of two spin-1/2 states giving a singlet, $j = 0$, or a triplet $j = 1$, state, so we write

$$\frac{1}{2} \otimes \frac{1}{2} = 0 \oplus 1. \tag{7.51}$$

The corresponding representations, $R(j_1)$ and $R(j_2)$, can be combined in the same way. More concretely, $R(j_1) \otimes R(j_2)$ can be decomposed as

$$R(j_1) \times R(j_2) = \sum_{j=|j_1-j_2|}^{j_1+j_2} R(j). \tag{7.52}$$

From the above discussion we see that arbitrarily large j can be obtained if we combine sufficiently many spins.

Not all values of j are relevant for the Chern–Simons theory. It turns out that for the expectation values of Wilson loops (7.45) the values of j higher than a certain $j_{\max} = k/2$ will give the same result as representations with $0 \leq j \leq j_{\max}$ (Elitzur *et al.*, 1989) (see Example III later). So we can write

$$0 \leq j \leq \frac{k}{2}, \tag{7.53}$$

where j can take all possible $k + 1$ integer and half-integer values, $0, 1/2, 1, \ldots, k/2$. As a result of this restriction the composition of representations can be expressed in terms of the non-equivalent ones in the following way (see Example III later):

$$R(j_1) \times R(j_2) = \sum_{j=|j_1-j_2|}^{b(j_1,j_2)} R(j), \tag{7.54}$$

where

$$b(j_1,j_2) = \min\{j_1 + j_2, k - j_1 - j_2\}. \tag{7.55}$$

This is a symbolic expression in the sense that it keeps from the tensor product of representations only the ones that give non-equivalent expectation values of the Wilson loops.

Relation (7.53) signifies that the SU(2) level-k theory has $k + 1$ distinct particles that correspond to different representations, $R(j)$, of the SU(2) group. The fusion rule between these anyonic particles is given by (7.54). This shows that the fusion of two anyons gives an outcome that can be one of the finitely many that belong to the anyonic model.

An anyon and its antiparticle fuse to the vacuum. For a given representation $R(j)$ that corresponds to an anyon there exists a dual representation $\bar{R}(j)$ such that their product gives the trivial representation with $j = 0$. In the anyonic language a source with the dual representation corresponds to the anti-anyon.

As an example, consider four sources in the fundamental representation of SU(2). We assume that the total charge of the sources is the trivial one, such that two of the representations are dual to the other two. Employing (7.54) and taking the four anyons R, R, $\bar{R}$ and $\bar{R}$, we have

$$R \times R = R_1 + R_2, \tag{7.56}$$

where R_1 and R_2 are the only two possible resulting representations of SU(2). This gives rise to a two-dimensional fusion space, which can encode one qubit.

7.2.3 The braiding evolution

The aim of this subsection is to determine the statistics of particles coupled to a non-Abelian Chern–Simons field. Our approach is rather different from the one we employed in the Abelian case. What we want to obtain is the unitary evolution that corresponds to the braiding of two sources. To do this we employ the Hamiltonian formalism of quantum field theory.

7.2.3.1 Hamiltonian approach

The starting point of a field theory is to write down an action (Hatfield, 1992; Sundermeyer, 1982; Weinberg, 1995). For illustration purposes let us take

$$I[\phi] = \int d^3x \left[G(x)\phi_0(x) + \pi_1(x)\dot{\phi}_1(x) - \mathcal{H}(x) \right], \tag{7.57}$$

where ϕ_0 and ϕ_1 are the fields of the theory and π_1 is the conjugate momentum of ϕ_1. We assume that there is no time derivative of ϕ_0, so there is no π_0 and the term $\mathcal{H}(x)$ depends only on ϕ_1 and π_1. In canonical quantisation we usually fix the timeslice at $t = 0$ with spatial coordinates $\mathbf{x} = (x^1, x^2)$. The quantisation condition requires that the field ϕ_1 and its canonical conjugate momentum π_1 are operators that satisfy

$$[\phi_1(\mathbf{x}), \pi_1(\mathbf{y})] = i\delta(\mathbf{x} - \mathbf{y}). \tag{7.58}$$

The states of the system are the wave functionals $\Psi[\phi_1]$ that formally satisfy the eigenvalue relation

$$H\Psi[\phi_1] = E\Psi[\phi_1], \tag{7.59}$$

where the Hamiltonian is given by $H = \int d^2\mathbf{x} \mathcal{H}(t = 0, \mathbf{x})$. This equation needs to be considered with caution as it is usually plagued with divergences that can be removed by appropriate renormalisation conditions (Symanzik, 1983). As the action does not involve a time derivative of ϕ_0, we can treat this variable as a Lagrange multiplier that enforces the constraint

$$G(\mathbf{x})\Psi[\phi_1] = 0. \tag{7.60}$$

The two relations (7.59) and (7.60) are sufficient to determine the wave functionals $\Psi[\phi_1]$ that describe the system and its properties.

This approach can be applied straightforwardly to the non-Abelian Chern–Simons theory. With a bit of algebra the action (7.35) can be brought to the form

$$I_{\text{CS}}[A] = \frac{k}{4\pi} \int d^3x \left[A_0^a \left(\partial_1 A_2^a - \partial_2 A_1^a - f^{abc} A_1^b A_2^c \right) + A_2^a \dot{A}_1^a \right]. \tag{7.61}$$

A comparison of this action with (7.57) dictates that the dynamical field is $A_1(\mathbf{x})$, with $A_2(\mathbf{x})$ being its conjugate momentum. Hence, we represent them with operators that satisfy

$$\left[A_1^a(\mathbf{x}), A_2^b(\mathbf{y}) \right] = i\delta^{ab} \frac{4\pi}{k} \delta(\mathbf{x} - \mathbf{y}). \tag{7.62}$$

Furthermore, A_0 is a Lagrange multiplier that imposes the Gauss law constraint. Comparing the actions (7.57) and (7.61) we also deduce that $H = 0$, identically. In other words, there is no dynamics in this model. The only degrees of freedom are gauge-related. As a result, the behaviour of the model is exclusively determined by the Gauss law constraint (7.60).

Similarly to the Abelian case, the sources in the Gauss law are introduced through the Wilson loops. Surprisingly, the sources are now matrices corresponding to the representation of the A field. Indeed, from (7.45) we see that the trace of the Wilson loop operators can be taken outside the functional integral. So the field A_0^a couples to a matrix. In the presence of two point-like sources at positions $\mathbf{x}_1$ and $\mathbf{x}_2$, the Gauss law, for the quantised theory, becomes

$$G^a(\mathbf{x})\Psi_{IJ}[A_1] = \frac{k}{4\pi} \left(\partial_1 A_2^a(\mathbf{x}) - \partial_2 A_1^a(\mathbf{x}) - f^{abc} A_1^b(\mathbf{x}) A_2^c(\mathbf{x}) \right) \Psi_{IJ}[A_1]$$
$$= -\delta(\mathbf{x} - \mathbf{x}_1)\Psi_{IK}[A_1](T_{(j)}^a)_{KJ} - \delta(\mathbf{x} - \mathbf{x}_2)\Psi_{IK}[A_1](T_{(j)}^a)_{KJ}, \tag{7.63}$$

where $(T_{(j)}^a)_{KJ}$ is the KJ element of the representation that corresponds to $\mathbf{x}_1$ and similarly for $\mathbf{x}_2$. In the presence of such sources the wave functional $\Psi_{IJ}[A_1]$ becomes a spinor where the first index, I, parameterises the components of the spinor and the second index, J, parameterises the different possible spinors.

Assume we know the wave functional $\Psi_0[A_1]$ that satisfies the Gauss law without sources, i.e., $G^a(\mathbf{x})\Psi_0[A_1] = 0$. Consider the Wilson line operator

$$U(C_{\mathbf{x}_1}) = \mathbf{P} \exp \left(i \int_{-\infty}^{x_1^1} ds\, T_{(j)}^a A_1^a(s, x_1^2) \right), \tag{7.64}$$

where $\mathbf{x}_1 = (x_1^1, x_1^2)$ and $C_{\mathbf{x}_1}$ is the path that goes from $x^1 = -\infty$ to $x^1 = x_1^1$, while the other coordinate is kept fixed at $x^2 = x_1^2$, as shown in Figure 7.4. With the help of (7.62) we can verify that

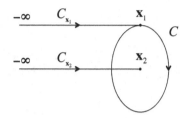

Fig. 7.4 Two sources are positioned at $\mathbf{x}_1$ and $\mathbf{x}_2$. They are described as the endpoints of two-dimensional Wilson lines, defined along the paths $C_{\mathbf{x}_1}$ and $C_{\mathbf{x}_2}$ starting at $-\infty$ and extending horizontally to the desired point. The source at $\mathbf{x}_1$ is braided around the source at $\mathbf{x}_2$ in a clockwise fashion, thus extending its path by the loop C.

$$\left[G^a(\mathbf{x}), U(C_{\mathbf{x}_1}) \right] = U(C_{\mathbf{x}_1}) T^a_{(j)} \delta(\mathbf{x} - \mathbf{x}_1). \tag{7.65}$$

Then the wave functional, which satisfies the Gauss law (7.63) with the two sources, is given by

$$\Psi_{IJ,MN}[A_1] = [U(C_{\mathbf{x}_1})]_{MI}[U(C_{\mathbf{x}_2})]_{NJ} \Psi_0[A_1], \tag{7.66}$$

where the M and N indices parameterise all possible solutions. This is equivalent to the wave function (2.5) of a charged particle in the presence of a magnetic field, or the quantum state (5.13) of the toric code in the presence of two sources. The wave functional $\Psi_{IJ,MN}[A_1]$ that satisfies the Gauss law is the complete solution as the Hamiltonian of the theory is identically zero. Hence, the eigenstate of the system in the presence of two independent sources at $\mathbf{x}_1$ and $\mathbf{x}_2$ is given in terms of two Wilson lines that connect infinity with the position of the sources, as shown in Figure 7.4.

7.2.3.2 Braiding matrix

To determine the statistics of this model we consider two sources described by the wave functional (7.66) given in terms of the Wilson lines (7.64). Then we braid source 1 around source 2 in a clockwise manner, as shown in Figure 7.4. The braiding evolution introduces a Wilson loop of the form

$$V(C) = \mathbf{P} \exp \left(i \oint_C d\mathbf{x} \cdot \mathbf{A}^a(\mathbf{x}) T^a_{(j)} \right), \tag{7.67}$$

that extends the Wilson line $U(C_{\mathbf{x}_1})$. The resulting wave functional is given by

$$\Psi'_{IJ,MN}[A_1] = [U(C_{\mathbf{x}_1})]_{MK}[V(C)]_{KI}[U(C_{\mathbf{x}_2})]_{NJ} \Psi_0[A_1], \tag{7.68}$$

where the product $U(C_{\mathbf{x}_1})V(C)$ gives the Wilson operator of the composite path $C_{\mathbf{x}_1} \cdot C$. To evaluate the relation between $\Psi_{IJ,MN}[A_1]$ and $\Psi'_{IJ,MN}[A_1]$ we do not need to know the exact form of $\Psi_0[A_1]$. We only need to know that it satisfies $V(C)\Psi_0[A_1] = \Psi_0[A_1]$. In other words, if we create two sources from the vacuum, move them along separate paths and then annihilate them, the resulting state is the vacuum. Hence, we just need to permute $V(C)$ and $U(C_{\mathbf{x}_2})$ to evaluate $\Psi'_{IJ,MN}[A_1]$. Note that if it were $[A^a_1(\mathbf{x}), A^b_2(\mathbf{y})] = 0$, then $[V(C), U(C_{\mathbf{x}_2})] = 0$. Indeed, the matrix representations of the two sources commute with each other as they act on different vector spaces. So the quantisation condition (7.62)

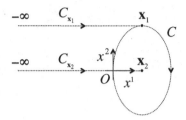

Fig. 7.5 The evolution given in Figure 7.4 is considered in discrete space. Possible non-trivial contributions in the permutation of the operators $[V(C)]_{KI}$ and $[U(C_{\mathbf{x}_2})]_{NJ}$ are from the intersection point O of paths C and $C_{\mathbf{x}_2}$. With a coordinate transformation we can set C to lie along x^2, while $C_{\mathbf{x}_2}$ remains along the coordinate x^1.

imposes a non-trivial permutation between these operators. The permutation step is very similar in spirit to the one we took to calculate the toric code braiding evolution in Subsection 5.2.1. A direct evaluation of this operation in the Chern–Simons case has been performed in Broda (1990); Guadagnini (1993). A simplified derivation is given below.

Consider the path configuration given in Figure 7.4. We want to permute operators $V(C)$ and $U(C_{\mathbf{x}_2})$. If the paths of these operators did not cross each other, then they would simply commute. From (7.62) we see that the non-trivial contribution in their permutation comes from the point at which the path C intersects $C_{\mathbf{x}_2}$. The intersection of these paths is called O in Figure 7.5. To facilitate the evaluation of the operator permutation we can discretise the Wilson operators by defining them on a square spatial lattice with infinitesimal lattice spacing. Then the paths are made out of consecutive links of the square lattice. The field $\mathbf{A}^a(\mathbf{i})$ is defined at the vertices of the lattice with coordinates $\mathbf{i} = (i^1, i^2)$, where i^1 is along x^1 and i^2 is along x^2. The field commutation relations in the discrete space become

$$[A_1^a(\mathbf{i}), A_2^b(\mathbf{j})] = i\delta^{ab}\frac{4\pi}{k}\delta_{\mathbf{ij}}. \tag{7.69}$$

In the discretised space the Wilson operators (7.64) and (7.67) can be written as a product of exponentials with each exponential corresponding to a lattice link that belongs to the Wilson line or loop. Such a discrete version of the Wilson line is given by

$$U(C_{\mathbf{x}_2}) = \mathbf{P}\exp\left(i\sum_{i^1} A_1^a(i^1, i^2)T_{(j)}^a\right) = \mathbf{P}\prod_{i^1}\exp\left(iA_1^a(i^1, i^2)T_{(j)}^a\right). \tag{7.70}$$

Here $A_1^a(i^1, i^2)$ is positioned at the starting point of each link as it is traversed by $C_{\mathbf{x}_2}$. From (7.69) we see that all of the exponentials of $[V(C)]_{KI}$ and $[U(C_{\mathbf{x}_2})]_{MK}$ commute with each other except when their paths cross at O. Then we are left to permute the two terms whose fields $\mathbf{A}^a(\mathbf{i})$ lie at the same point O, one from the Wilson line (7.64) and one from the Wilson loop (7.67).

We now focus our analysis on the point O. As the system is invariant under coordinate transformations, we can choose the coordinates at O so that the path C meets the path $C_{\mathbf{x}_2}$ vertically. That is, the tangent of C at O is along x^2, while $C_{\mathbf{x}_2}$ remains along x^1, as shown in Figure 7.5. The non-commuting parts of the exponentials from the two Wilson operators are given by $\exp\left(iA_1^a(\mathbf{i}_O)T_{(j)}^a\right)$ and $\exp\left(iA_2^b(\mathbf{i}_O)T_{(j)}^b\right)$. To proceed we employ the following property (see Exercise 7.3). For two operators $\mathcal{A}$ and $\mathcal{B}$ with a commutator $[\mathcal{A}, \mathcal{B}]$ that commutes with both $\mathcal{A}$ and $\mathcal{B}$ (i.e., $[\mathcal{A}, [\mathcal{A}, \mathcal{B}]] = [\mathcal{B}, [\mathcal{A}, \mathcal{B}]] = 0$) we have

$$e^{\mathcal{A}} e^{\mathcal{B}} = e^{[\mathcal{A}, \mathcal{B}]} e^{\mathcal{B}} e^{\mathcal{A}}. \tag{7.71}$$

This allows us to reverse the order of $[V(C)]_{KI}$ and $[U(C_{\mathbf{x}_2})]_{NJ}$ up to a multiplicative term $e^{[\mathcal{A}, \mathcal{B}]}$. To calculate it we employ (7.69) and set $\mathcal{A} = iA_2^a(\mathbf{i}_O)T_{(j)}^a$ and $\mathcal{B} = iA_1^b(\mathbf{i}_O)T_{(j)}^b$, finally giving

$$[\mathcal{A}, \mathcal{B}] = -[A_2^a(\mathbf{i}_O), A_1^b(\mathbf{i}_O)]T_{(j)}^a \otimes T_{(j)}^b = i\frac{4\pi}{k}T_{(j)}^a \otimes T_{(j)}^a. \tag{7.72}$$

Hence, we can relate the states that describe the braided and unbraided configurations by

$$\Psi'_{IJ,MN}[A_1] = B_{M,N}^{X,Y}\Psi_{IJ,SQ}[A_1], \tag{7.73}$$

where the X, M indices parameterise point 1 and Y, N parameterise point 2. The braiding matrix is given by

$$B = q^{T_{(j)}^a \otimes T_{(j)}^a}, \tag{7.74}$$

for

$$q = \exp\left(i\frac{4\pi}{k}\right). \tag{7.75}$$

The braiding matrix B remains the same in the continuum limit as it does not depend on the lattice spacing. This can also be shown with a more involved calculation performed in the continuum. Note that the lattice discretisation is just a technical step employed to evaluate the permutation of the Wilson operators. It is not related to defining Chern–Simons theories on a lattice. Nevertheless, (7.75) is a semiclassical result that holds in the large k limit. The exact renormalisation treatment, corresponds to substituting k with $k + 2$ for the SU(2) group (Witten, 1989). Hence, the exact result for all values of k is given by

$$q_{\text{exact}} = \exp\left(i\frac{4\pi}{k + 2}\right), \tag{7.76}$$

which tends to (7.75) when k is taken to be large. Relation (7.74) also holds when the two sources have different representations. The braiding matrix, B, describes the non-trivial effect which two successive exchanges between two sources can have on their quantum state. Hence, the Chern–Simons theories can support particles with non-Abelian statistics. Interestingly, this relation applies to the Abelian case as well. It is known (Witten, 1989) that the large k limit of the non-Abelian theory is closely related to the Abelian one. Indeed, when we employ (7.73), (7.74) and (7.75) for the case of two Abelian sources with charge Q we obtain (7.13).

Consider the case where the employed group is SU(2). It has been shown that when we use this anyonic system with the qubit encoding given in (7.56) the braiding matrices (7.74) with (7.76) can give rise to universal quantum computation for $k = 3$ or $k > 5$ (Freedman *et al.*, 2002b). This result becomes all the more important as Chern–Simons theories are expected to describe the physics of the fractional quantum Hall effect.

7.3 Example I: Braiding for the SU(2) Chern–Simons theory

In this example we analyse the braiding properties of the SU(2) Chern–Simons theory level k in the fundamental representation, $j = 1/2$. In this case $Q = 1$ and $T^a = \sigma^a/2$. From (7.74) and (7.76) we have the braiding matrix

$$B = \exp\left(i\frac{4\pi}{k+2}\frac{\sigma^a}{2} \otimes \frac{\sigma^a}{2}\right), \tag{7.77}$$

where there is a summation over $a = 1, 2, 3$. This matrix acts on a two spin-1/2 space of states, parameterised by

$$|\Psi\rangle = \sum_{i,j=0,1} c_{ij} |ij\rangle \tag{7.78}$$

with complex numbers c_{ij} satisfying $\sum_{i,j} |c_{ij}|^2 = 1$. We can derive that

$$\sum_{a=1}^{3} \sigma^a \otimes \sigma^a = -\begin{pmatrix} 1 & 0 & 0 & 0 \\ 0 & 1 & 0 & 0 \\ 0 & 0 & 1 & 0 \\ 0 & 0 & 0 & 1 \end{pmatrix} + 2\begin{pmatrix} 1 & 0 & 0 & 0 \\ 0 & 0 & 1 & 0 \\ 0 & 1 & 0 & 0 \\ 0 & 0 & 0 & 1 \end{pmatrix} = -\mathbb{1} + 2E, \tag{7.79}$$

with $\mathbb{1} = \delta_{ij}\delta_{kl}$ and $E = \delta_{il}\delta_{kj}$, where i, j are the indices of the operator acting on the first spin and k, l are the indices acting on the second. Hence, this braiding operator can be written as

$$B = \exp\left[-i\frac{\pi}{k+2}(\mathbb{1} - 2E)\right] = \exp\left(-i\frac{\pi}{k+2}\right)\left(\mathbb{1}\cos\frac{2\pi}{k+2} + iE\sin\frac{2\pi}{k+2}\right), \tag{7.80}$$

as $E^2 = 1$. For $k = 2$ we obtain the non-trivial evolution

$$B = e^{i\frac{\pi}{4}}E. \tag{7.81}$$

This matrix interchanges the states of the two spins as $E|ij\rangle = |ji\rangle$. The resulting braiding matrix differs from the Ising model matrix (6.58) we derived in the honeycomb lattice case by an overall phase factor $e^{-i\pi/2}$ (Rowell et al., 2009).

7.4 Example II: From bulk to boundary

The Chern–Simons action is defined on three-dimensional spacetime M. We now show that this action can be written in terms of a two-dimensional integral on the boundary ∂M. This property is not just a mathematical nicety. It is closely related to the invariance of the action under three-dimensional coordinate transformations that gives rise to topologically invariant observables. In other words, if the information of a system can be efficiently encoded on its boundary then there is a lot of redundant information in the coordinate parameterisation of the bulk. This is expressed as an invariance of the system's observables under coordinate transformations.

7.4.1 Abelian case

Consider the action of the Abelian Chern–Simons theory defined on the three-dimensional space M given by

$$I_{CS}[A] = \frac{m}{2} \int_M d^3x\, \varepsilon^{\mu\nu\rho} A_\mu \partial_\nu A_\rho. \tag{7.82}$$

We want to demonstrate the topological properties of this action by showing that it can be reduced to a boundary term. Let us take the Clebsch decomposition of the gauge field, given by

$$A_\mu = \partial_\mu \theta + \alpha \partial_\mu \beta. \tag{7.83}$$

This provides a general reparameterisation of A_0, A_1 and A_2 in terms of the fields θ, α and β. Substituting in the action $I_{CS}[A]$ we obtain

$$I_{CS}[A] = \frac{m}{2} \int_{\partial M} dS^\mu \varepsilon^{\mu\nu\rho} \theta \partial_\nu \alpha \partial_\rho \beta, \tag{7.84}$$

defined exclusively on the boundary ∂M of M. Hence, deformations of the field A_μ are considered to be equivalent if they do not alter its value at the boundary, where the topology of the system is encoded.

7.4.2 Non-Abelian case

The non-Abelian Chern–Simons action is given by

$$I_{CS}[A] = \frac{k}{4\pi} \int_M d^3x\, \varepsilon^{\mu\nu\rho}\, \mathrm{tr}\big(A_\mu \partial_\nu A_\rho + i\frac{2}{3}A_\mu A_\nu A_\rho\big), \tag{7.85}$$

where we take M to be the infinite three-dimensional space $\mathbb{R}^3$. In the absence of sources the field is a pure gauge, $A_\mu = -iU^\dagger \partial_\mu U$. Then the action becomes $I_{CS}[A] = 2\pi\omega[U]$ with

$$\omega[U] = \frac{1}{24\pi^2} \int_M d^3x\, \varepsilon^{\mu\nu\rho} \mathrm{tr}\big(U^\dagger \partial_\mu U U^\dagger \partial_\nu U U^\dagger \partial_\rho U\big), \tag{7.86}$$

the winding number we met also in (7.38). We now demonstrate explicitly that $\omega[U]$ is an integer. For convenience we take $U \in SU(2)$ with the explicit form

$$U = \exp\left(i\lambda^a \frac{\sigma^a}{2}\right), \tag{7.87}$$

where σ^a are the Pauli matrices and $\lambda^a(x)$ are general real functions of $x \in M$. We can employ the property

$$\exp\left(i|\lambda|\hat{\lambda} \cdot \boldsymbol{\sigma}\right) = \mathbb{1}\cos|\lambda| + i\hat{\lambda} \cdot \boldsymbol{\sigma}\sin|\lambda| \tag{7.88}$$

and substitute U in $\omega[U]$ to finally find

$$\omega[U] = \frac{1}{16\pi^2}\int_{\partial M} dS^\mu \varepsilon^{\mu\nu\rho}\varepsilon_{abc}\hat{\lambda}^a\partial_\nu\hat{\lambda}^b\partial_\rho\hat{\lambda}^c(|\lambda| - \sin|\lambda|), \tag{7.89}$$

where $|\lambda| = \sqrt{\lambda^a\lambda^a}$ and $\hat{\lambda}^a = \lambda^a/|\lambda|$ for $a = 1, 2, 3$. So, with this parameterisation we managed to rewrite the non-Abelian Chern–Simons action as a surface boundary integral.

We can choose now the behaviour of λ at infinity to be

$$|\lambda(r \to \infty)| \to 2\pi n, \tag{7.90}$$

so that

$$U(r \to \infty) \to \pm\mathbb{1}. \tag{7.91}$$

This is a boundary condition of the group elements U, while the behaviour of U can be arbitrary everywhere else. This gives

$$\omega[U] = \frac{n}{8\pi}\int_{\partial M} dS^\mu \varepsilon^{\mu\nu\rho}\varepsilon_{abc}\hat{\lambda}^a\partial_\nu\hat{\lambda}^b\partial_\rho\hat{\lambda}^c. \tag{7.92}$$

We previously encountered a similar integral in the definition of the Chern number (6.49). There we saw that it gives the winding number of the mapping from the surface ∂M to the two-dimensional sphere parameterised by $\hat{\lambda}^a$. From (7.90) we see that $\hat{\lambda}^a$ takes non-trivial values over a wide range, while the integral in (7.92) corresponds to a non-trivial unit winding. This gives

$$\omega[U] = n. \tag{7.93}$$

In other words, $\omega[U]$ depends only on the global properties of U. Equivalently, the action $I_{CS}[A]$ depends on the global properties of A_μ. Note that (7.92) is proportional to (7.84) of the Abelian case.

7.5 Example III: Non-Abelian anyons and their fusion rules

Here we demonstrate (7.53). This relation dictates that there is only a finite number of different species of anyonic sources emerging from the non-Abelian Chern–Simons theory with level k. We also demonstrate the fusion rules of these anyons. For these tasks we follow the methodology given in Elitzur *et al.* (1989). In this reference and in what follows the Gauss law is solved at the classical level. The resulting vector potential can then be employed in the functional integral (7.45) for evaluating the expectation value of arbitrary Wilson loop operators.

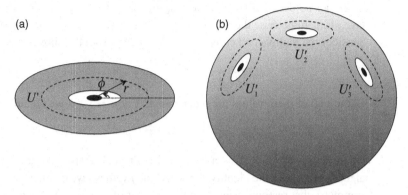

Fig. 7.6 (a) An annulus is depicted with surface parameterised by polar coordinates $r_1 \leq r \leq r_2$ and $0 \leq \phi \leq 2\pi$. A source is placed at the centre of the puncture depicted as a black disk. The presence of the source is witnessed by the holonomy U' with a loop that wraps around the puncture. (b) Consider the surface of a sphere with three punctures each supporting a source. These sources have total trivial charge, i.e., $U'_1 U'_2 U'_3 = \mathbb{1}$. This can be verified by taking a loop around all three punctures that corresponds to the holonomy of the product $U'_1 U'_2 U'_3$. As the loop is contractible due to the compact topology of the sphere, it should correspond to the trivial holonomy $\mathbb{1}$.

7.5.1 Number of anyonic species

Consider the SU(2) Chern–Simons theory defined on an annulus. In polar coordinates (r, ϕ) the surface of the annulus is parameterised by $r_1 \leq r \leq r_2$ and $0 \leq \phi \leq 2\pi$. The puncture at the origin of the disk can harbour a source described by the representation λ, as shown in Figure 7.6(a). For convenience, we solve the Gauss law in terms of the appropriate vector potential. As there are no sources directly on the annulus, the vector potential is a pure gauge given by

$$A_\mu = -iU^\dagger \partial_\mu U. \tag{7.94}$$

We want this vector potential to satisfy the Gauss law (7.63), but with one source. For that we take U to be an SU(2) element parameterised by (Elitzur *et al.*, 1989)

$$U(r, \phi, t) = \tilde{U}(r, \phi, t) U'(\phi, t) \text{ with } U'(\phi, t) = e^{i\frac{\phi}{k}\lambda(t)}. \tag{7.95}$$

The unitary $\tilde{U}$ is an arbitrary group element that is single-valued on the annulus, i.e., $\tilde{U}(r, 2\pi, t) = \tilde{U}(r, 0, t)$. The unitary $U'(\phi, t)$ arises from the presence of a source at the origin. As shown in Figure 7.6(a), $U'(2\pi, t)$ corresponds to a Wilson loop (or holonomy) with respect to a loop that wraps around the source. For $k > 1$ it is multivalued as $U'(2\pi, t) \neq U'(0, t) = \mathbb{1}$. The Abelian version of this is the Aharonov–Bohm effect, where a non-trivial phase factor emerges when a charge circulates a flux.

It is clear that the system is invariant under gauge transformations that act by multiplying the element U of (7.95) both from the left and from the right. We want to observe the behaviour of U' when the source λ is in a particular j representation of the SU(2) group. By a gauge transformation we can orient λ along the third direction in SU(2) space, i.e., $\lambda = 2jT^3$, where j can be an integer or a half-integer. The integral value of $2j$ can be deduced by a similar argument to the Abelian case (Yang, 1970). We can take j to parameterise the different types of sources. To study the presence of a source we consider the multivalued component at $\phi = 2\pi$. From the above analysis we have

$$U'(2\pi, t) = e^{i\frac{4\pi j}{k}T^3}, \text{ for } j = 0, \pm\frac{1}{2}, \pm1, \pm\frac{3}{2}, \pm2, \ldots \tag{7.96}$$

But not all of these values are independent. For $j' = j + k$ we obtain

$$e^{i\frac{4\pi j'}{k}T^3} = e^{i\frac{4\pi j}{k}T^3} e^{i4\pi T^3} = e^{i\frac{4\pi j}{k}T^3}, \tag{7.97}$$

where $\exp(i4\pi T^3) = 1$ as a rotation by 4π with respect to any direction of the su(2) algebra gives back the identity. So not all the j values give distinct sources. There is another symmetry that restricts even more the number of sources. With a gauge transformation by $U_1 = \exp(i\pi T^1)$, we can rotate U' around the 1 axis so that

$$U_1 U'(2\pi, t) U_1^\dagger = e^{i\frac{4\pi(-j)}{k}T^3}. \tag{7.98}$$

As a result, the substitution $j' = -j$ gives the same type of sources, as their effect is related by a gauge transformation. The identification of the representations under the transformations

$$j' = j + k \text{ and } j' = -j \tag{7.99}$$

reduces the number of independent sources to $k + 1$ many. For convenience we choose the sources to be parameterised by

$$j = 0, \frac{1}{2}, 1, \ldots, \frac{k}{2}. \tag{7.100}$$

Hence, there are $k + 1$ different anyonic sources emerging from the SU(2) level-k Chern–Simons theory.

7.5.2 Fusion rules

To study the fusion rules between the sources of the SU(2) level-k Chern–Simons theory we consider the case where there are three sources present in their corresponding punctures on the surface of a sphere, as shown in Figure 7.6(b). Each source is described by its holonomy

$$U'_i = e^{i\frac{4\pi j_i}{k}T^3} \text{ for } i = 1, 2, 3. \tag{7.101}$$

The holonomy given by the product of these three holonomies, $U'_1 U'_2 U'_3$, corresponds to a loop that wraps around all three sources. Due to the geometry of the sphere it can be contracted to a point, giving rise to the trivial holonomy. This statement takes the following form:

$$e^{i\frac{4\pi j_1}{k}T^3} e^{i\frac{4\pi j_2}{k}T^3} e^{i\frac{4\pi j_3}{k}T^3} = \mathbb{1}. \tag{7.102}$$

Importantly, (7.102) holds for all equivalent representations that can be obtained with the transformations (7.99). When $j_1 + j_2 \leq k/2$ then from (7.50) we have that $|j_1 - j_2| \leq j_3 \leq j_1 + j_2$, which follows the standard composition rules of the SU(2) representations. This range of j_3 parameters corresponds to independent sources as they are all smaller than $k/2$. On the other hand, when $j_1 + j_2 > k/2$ we can rewrite j_1 and j_2 in (7.102) so that j_3 takes values smaller than $k/2$ as well. Let us multiply the right-hand side with two elements $e^{i2\pi T^3} = \pm 1$, where the sign corresponds to integer or half-integer representations of T^3. In the case of three sources, either two of them are half-integer and the third integer or they all have integer j. We can combine two $e^{i2\pi T^3}$ terms with the possible half-integer j's, say j_1 and j_2, without changing the right-hand side of (7.102). By employing the (7.99) symmetries we can change the sign of j_1 and j_2 as well. With these steps relation (7.102) becomes

$$e^{i\frac{4\pi}{k}(\frac{k}{2}-j_1)T^3} e^{i\frac{4\pi}{k}(\frac{k}{2}-j_2)T^3} e^{i\frac{4\pi j_3}{k}T^3} = \mathbb{1}, \qquad (7.103)$$

which effectively substitutes j_1 with $k/2 - j_1$ and j_2 with $k/2 - j_2$. As a conclusion, the upper bound of the possible values of j_3 becomes $j_3 \leq (k/2 - j_1) + (k/2 - j_2)$ or in other words

$$j_3 \leq k - (j_1 + j_2). \qquad (7.104)$$

With this condition the value of j_3 satisfies $j_3 < k/2$ as we have $(k/2 - j_1) + (k/2 - j_2) < k/2$. If instead j_1 and j_3 were half-integers then similar manipulations of (7.102) would produce the same condition (7.104). This derives the result given in (7.55).

Summary

In this chapter we studied the Chern–Simons theories. These topological quantum field theories can describe general Abelian anyons as well as a particular class of non-Abelian anyons. The Abelian Chern–Simons action emerges as part of electromagnetism when we restrict to two spatial dimensions. What this action does is make any direct interactions between sources short-ranged. At the same time it allows topological interactions of the Aharonov–Bohm type that give rise to anyonic statistics. Non-Abelian Chern–Simons theories support a discrete family of non-Abelian anyons. This family has a variety of fusion properties and statistical behaviours that can be described with general formulae, such as the braiding matrix (7.74). For more details we refer to more specialised literature (Dunne, 1998; Wang, 2010).

We also studied the expectation value of Wilson loop operators, $\langle W(L) \rangle$. The loops of these operators can be considered as the worldlines of anyons that can form any general link, L. Due to coordinate transformation invariance of the action and of the Wilson loop operators, the expectation value $\langle W(L) \rangle$ is invariant under any continuous deformation of the link L. Hence, Chern–Simons theories give rise to topological invariants of links. In the case of Abelian Chern–Simons theories such an invariant is given in the form of the Gauss integral. The latter corresponds to the linking or winding number between two strands. This was our first encounter of the relation between Chern–Simons theories and topological invariants of links. While two links with different linking numbers are necessarily non-equivalent, there is a large class of links with the same linking number that are not equivalent under continuous deformations of the links. The generalisation to non-Abelian Chern–Simons theories provides more sophisticated tools to address the question of distinguishability between two links. They give rise to the Jones polynomial that is presented in the next chapter.

Exercises

7.1 Demonstrate that the Abelian Chern–Simons action (7.7) is invariant under the gauge transformation

$$A_\mu(x) \rightarrow A_\mu^\omega(x) = A_\mu(x) + \partial_\mu \omega(x). \tag{7.105}$$

For the derivation assume that A_μ goes to zero at the boundaries of the system.

7.2 Consider the Abelian Chern–Simons theory for a given m. Take a source to be a disk of radius R with homogeneous distribution of charge Q on it. Derive the spin (7.14) of the source by calculating the total phase accumulated from the charge when we rotate the disk by 2π. [*Hint*: Evaluate the phase of a small ring of charge of radius $0 \leq r \leq R$ due to the flux enclosed by the ring.]

7.3 Prove that

$$e^{-\lambda \mathcal{B}} \mathcal{A} e^{\lambda \mathcal{B}} = \mathcal{A} + \lambda [\mathcal{A}, \mathcal{B}], \tag{7.106}$$

when $[\mathcal{A}, [\mathcal{A}, \mathcal{B}]] = [\mathcal{B}, [\mathcal{A}, \mathcal{B}]] = 0$. Then demonstrate that relation (7.71) holds. [*Hint*: Consider the function $f(x) = e^{x\mathcal{A}} e^{x\mathcal{B}}$ and its differentiation with respect to x.]

PART III

QUANTUM INFORMATION
PERSPECTIVES

The study of anyonic systems as computational means has led to the exciting discovery of a new quantum algorithm. This algorithm provides a novel paradigm that fundamentally differs from searching (Grover, 1996) and factoring (Shor, 1997) algorithms. It is based on the particular behaviour of anyons and its goal is to evaluate Jones polynomials (Jones, 1985, 2005). These polynomials are topological invariants of knots and links, i.e., they depend on the global characteristics of their strands and not on their local geometry. The Jones polynomials were first connected to topological quantum field theories by Edward Witten (1989). Since then they have found applications in various areas of research, such as biology for DNA reconstruction (Nechaev, 1996) and statistical physics (Kauffman, 1991).

The best known classical algorithm for the exact evaluation of the Jones polynomial demands exponential resources (Jaeger *et al.*, 1990). Employing anyons involves only a polynomial number of resources to produce an approximate answer to this problem (Freedman *et al.*, 2003b). Evaluating Jones polynomials by manipulating anyons resembles an analogue computer. Indeed, the idea is equivalent to the classical setup, where a wire is wrapped several times around a solenoid that confines magnetic flux. By measuring the current that runs through the wire one can obtain the number of times the wire was wrapped around the solenoid, i.e., their linking number. Similarly, by creating anyons and spanning links with their worldlines we are able to extract the Jones polynomials of these links (Kauffman and Lomanaco, 2006). The translation of this anyonic evolution to a circuit-based quantum algorithm was demonstrated explicitly in (Aharonov *et al.*, 2009).

Jones polynomials have been introduced to establish equivalences between knots or links. A knot is a single component strand living in a three-dimensional space that has no open ends. A link consists of many strand components with no open ends. For simplicity, we refer in the following to a link also in the single component case. The question we want to address is whether two links are equivalent or not. The equivalence is with respect to isotopy moves, i.e., any kind of continuous deformations of the link components apart from cutting them open and reconnecting them. One can check that the problem of equivalence of links becomes intractable as the size of the links, measured by the number of twists and crossings, increases.

It is a rather surprising fact that the equivalence relations between links can be systematically characterised in a mathematical way. Based on this characterisation it is possible to construct link invariants, such as the Jones polynomials, that do not change under isotopy moves. Hence, if two links have different link invariants then it is not possible to continuously deform one of them into the other. In other words, they cannot be topologically equivalent. This is a fascinating property which allows the systematic study of links.

In the following sections we present the isotopy moves that characterise the equivalencies between links. With simple steps we then introduce the Jones polynomial of a link, that remains unchanged under the application of isotopy moves. This method is based on geometrical considerations. It demonstrates that equivalent links necessarily have the same Jones polynomial. To complement it we develop an algebraic approach based on a particular representation of braids. This method provides the means to evaluate Jones polynomials in terms of anyonic evolutions.

8.1 From link invariance to Jones polynomials

The quantum properties of non-Abelian anyons can be given by the expectation value of their spacetime evolutions. Suppose we generate anyons from the vacuum, braid them and then ask for the probability that they fuse back to the vacuum. To find this probability we need to calculate the expectation value of the anyonic evolution with worldline trajectories that form closed paths, i.e., links. Due to the statistical nature of the anyonic evolutions, these expectation values are invariant under continuous deformations of trajectories. Hence, they are topological link invariants. Such expectation values were calculated in the previous chapter. There, we mentioned that Jones polynomials appear as the expectation values of anyonic evolutions described by non-Abelian Chern–Simons theories (Witten, 1989).

Here, we explicitly construct the Jones polynomials by studying the topological invariance of links. Initially, we define equivalency classes between links that can be continuously deformed into each other. Three elementary isotopy moves, called the Reidemeister moves, are sufficient to establish these topological equivalencies. Next, we assign complex numbers to topological properties of the links, such as the braiding of their strands. The final outcomes are quantities that are invariant under any continuous deformation of the links. Still these quantities keep information about their topological characteristics. These are the topological link invariants we are looking for.

8.1.1 Reidemeister moves

To rigorously define the set of continuous deformations of a given link we need to first systematically categorise its geometrical properties. To achieve that, we project the link onto a plane. This gives a two-dimensional diagram, where we keep the overpasses and underpasses of the strands intact according to their natural three-dimensional configuration. This is a necessary step to avoid the dependence of the crossings on our perspective of the link. Such a planar projection can be achieved by continuous deformations of the original link.

Now we can present the Reidemeister theorem (Alexander and Briggs, 1926; Reidemeister, 1926). It states that two links can be deformed continuously from one into the other if

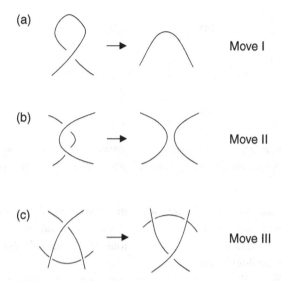

Fig. 8.1   The Reidemeister moves that relate equivalent links. (a) Move I undoes a twist of a strand. (b) Move II separates two unbraided strands. (c) Move III slides a strand under a crossing.

and only if their two-dimensional projections can be deformed from one to the other via a sequence of three simple local moves. These are presented in Figure 8.1 and concern parts of the links. They can be applied iteratively as many times as deemed necessary. In particular, move I undoes a twist. This move leaves the link invariant as it is not involved with a self-knotting or braiding with another link. Move II separates two strands of the link if they are not braided together. Finally, move III concerns three strands of a link. It allows us to slide a strand under the crossing of two other strands. All these isotopy moves correspond to continuous deformations of strands, which can define topological equivalences between links. In other words, Reidemeister moves can transform between two-dimensional projections of any two equivalent links.

8.1.2   Skein relations and Kauffman brackets

After identifying the equivalences between links, we would like to associate numbers to them. In particular, we introduce a polynomial denoted as $\langle L \rangle$, where L can be a link or parts of it. This polynomial, known as the state sum or Kauffman bracket, was initially introduced by Louis Kauffman (1991). The state sum is defined to be invariant under isotopy moves that transform links in a continuous way. Our strategy for the evaluation of $\langle L \rangle$ for a given link is to decompose it into the state sum of simpler links. This can be done with certain steps that keep information of the decomposition process.

The topologically simplest link configuration comprises disentangled loops. We use this as the reference configuration. Our aim is to find a process that relates the state sum of a general link to the state sum of disentangled ones. For that we introduce the Skein relations that split the crossings, as shown in Figure 8.2. The Skein relations replace the state sum of

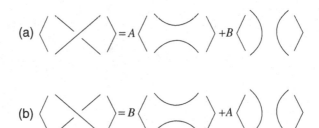

Fig. 8.2 Splitting an actual crossing according to its orientation, (a) or (b), by substituting it with two other avoiding crossings weighted with complex numbers A and B. By rotating all of the crossings in (a) by $\pi/2$ we obtain relation (b).

an actual crossing with the state sum of two avoided crossings. They also introduce general complex numbers A and B as weights to horizontal and vertical splittings of a crossing. In a sense, these are the defining relations that produce the Jones polynomials instead of some other topological invariants.

When a Skein relation is applied to a given crossing it produces two new graphs, with the rest of the graph remaining the same. If we apply it on all N actual crossings of a graph we obtain the sum of 2^N different elementary graphs. The final configuration is a sum of disjoint, unentangled loop configurations denoted by S and weighted by products of A's and B's. We denote the number of these loops at each configuration S by $|S|$. Moreover, we define the following properties of the state sum with respect to loops

$$\langle K \cup \bigcirc \rangle = d\langle K \rangle, \tag{8.1}$$

where K is a non-empty link, $\bigcirc$ is a separate loop and d is a real constant. For the case of an isolated loop, we define

$$\langle \bigcirc \rangle = 1. \tag{8.2}$$

The Skein relations of Figure 8.2 together with (8.1) and (8.2) uniquely define the state sum of a link L. Its general form is given by

$$\langle L \rangle = \sum_{\{S\}} d^{|S|-1} A^i B^j. \tag{8.3}$$

The summation runs over all 2^N possible configurations S, which result from the splitting of the graph. Also, i and $j = N - i$ are the times a horizontal or a vertical splitting, respectively, was employed in order to obtain the S configuration from the initial link L. Hence, we have assigned a number to a link in the form of the state sum.

Our task now is to choose A, B and d so that the state sum $\langle L \rangle$ is invariant under the Reidemeister moves II and III. Later on we shall see how to decorate $\langle L \rangle$ to make it invariant with respect to the Reidemeister move I as well. We initially apply the Skein relation to the state sum of the diagram in the left-hand side of Reidemeister move II. The resulting configuration is given in Figure 8.3. It implies that the state sum is invariant with respect to the Reidemeister move II if

$$ABd + A^2 + B^2 = 0 \quad \text{and} \quad AB = 1. \tag{8.4}$$

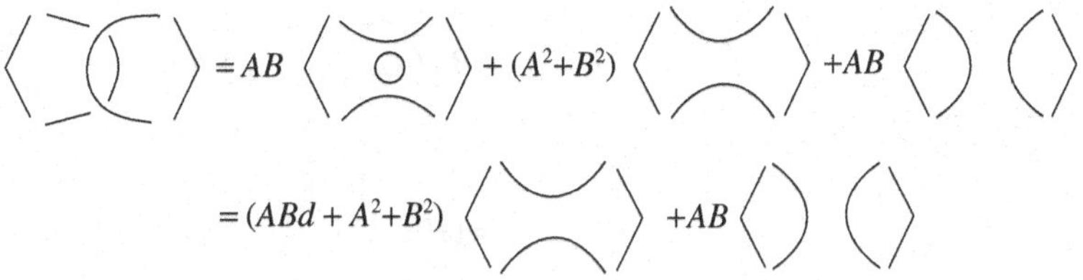

$$\left\langle \text{⟨)} \right\rangle = AB \left\langle \bigcirc \right\rangle + (A^2 + B^2) \left\langle \text{⟩⟨} \right\rangle + AB \left\langle \, \right\rangle \left\langle \, \right\rangle$$

$$= (ABd + A^2 + B^2) \left\langle \text{⟩⟨} \right\rangle + AB \left\langle \, \right\rangle \left\langle \, \right\rangle$$

Fig. 8.3 Applying the Skein relations to the state sum of the diagram on the left-hand side of the Reidemeister move II produces two new graphs. To make them equal to the right-hand side of this move we set $ABd + A^2 + B^2 = 0$ and $AB = 1$.

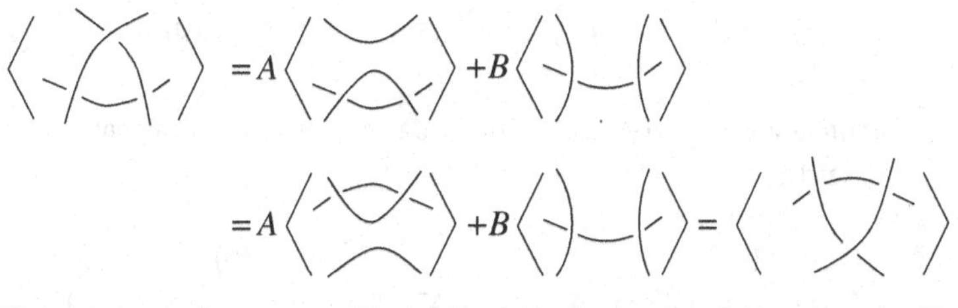

$$\left\langle \, \right\rangle = A \left\langle \, \right\rangle + B \left\langle \, \right\rangle$$

$$= A \left\langle \, \right\rangle + B \left\langle \, \right\rangle = \left\langle \, \right\rangle$$

Fig. 8.4 Applying the Skein relations to the state sum of the left-hand side diagram of the Reidemeister move III automatically gives the right-hand side. In the second step we employed the Reidemeister move II holding for the B and d given by (8.5).

These relations determine both B and d in terms of A as follows:

$$B = A^{-1}, \quad d = -A^2 - A^{-2}. \tag{8.5}$$

Next we consider the Reidemeister move III. Figure 8.4 shows that no extra condition between A, B and d needs to be imposed in order to satisfy it.

So the state sum $\langle L \rangle$ is in general a Laurent polynomial in A, given by

$$\langle L \rangle = \sum_{\{S\}} d^{|S|-1} A^{i-j} \quad \text{with } d = -A^2 - A^{-2}. \tag{8.6}$$

The evaluation of the state sum for various simple links is given in Example I later in this chapter. To create a polynomial that also satisfies the Reidemeister move I we need to introduce the Jones polynomials.

8.1.3 Jones polynomial

We can directly verify that the state sum is in general not invariant under Reidemeister move I. Consider the state sum of a single twist such as the one presented in Figure 8.5. Applying the Skein relations we see that the state sum introduces an extra factor $(-A)^3$. Since this is an overall factor, it can easily be accounted for. A new polynomial, the Jones polynomial, that is invariant under all Reidemeister moves for any A is defined by

$$\left\langle \ \asymp \ \right\rangle = B \left\langle \ \smile\frown \ \right\rangle + A \left\langle \ \bigcirc \ \right| \right\rangle$$

$$= (B + dA) \left\langle \ \big) \big(\ \right\rangle = (-A)^3 \left\langle \ \big) \big(\ \right\rangle$$

Applying the Skein relations to the state sum of Reidemeister move I produces an extra factor $(-A)^3$. Hence, the state sum is not invariant under twists of strands.

(a) i $\times$ $w_i = +1$ (b) i $\times$ $w_i = -1$

The writhe $w(L)$ of an oriented link L is defined as the sum of the signs assigned to each oriented crossing of L as given in (a) and (b).

(a) $\times = A \ \smile\frown \ + A^{-1} \ \big) \big($ (b) $\times = A^{-1} \ \smile\frown \ + A \ \big) \big($

(c) $\bigcirc = -A^2 - A^{-2} = d$

The reduction procedures applied to the state sum of a link to give finally a polynomial in A. (a) and (b) depict the Skein relations, while (c) assigns the number d to every loop.

$$V_L(A) = (-A)^{3w(L)} \langle L \rangle, \tag{8.7}$$

where $w(L)$ is the writhe or twisting of the link L. To determine the writhe we need to give an orientation to all link components. Then we take the sum of signs for all crossings

$$w(L) = \sum_i w_i, \tag{8.8}$$

where the w_i's are assigned to each vertex i as in Figure 8.6. The writhe is well defined as it does not depend on the choice of initial orientations of the link. Whenever a twist gives rise to a factor $(-A)^{-3}$, such as the one in Figure 8.5, the prefactor $(-A)^{3w(L)}$ contributes $(-A)^3$ and cancels it. The Jones polynomial $V_L(A)$ is hence insensitive to all Reidemeister moves.

In Figure 8.7 we summarise the Skein relations that are compatible with the Reidemeister moves and the contributions from an independent loop. For $A = t^{-1/4}$, the polynomial $V_L(A)$ agrees with the original definition of the Jones polynomial. The above construction is a diagrammatic derivation of the Jones polynomials given by Kauffman (1991). The motivation was the construction of topological invariants of links. In the next section we present an algebraic derivation of the Jones polynomials based on the properties of the braid group.

8.2 From the braid group to Jones polynomials

We now take an algebraic approach to derive Jones polynomials that is closer to the original method Vaughan Jones used when he discovered them (Jones, 1985). This approach is based on the braid group. The braid group can be employed to describe the statistics between anyons. It only accounts for the topological properties of anyonic worldlines and is independent of their geometrical characteristics. So this braid group approach is expected to also give rise to topological invariants. In the following we present the braid group and its properties. Then we define a trace of the braid group elements and derive the Jones polynomials.

8.2.1 The braid group

Let us denote by b_i for $i = 1, \ldots, n-1$ the generators of the braid group B_n. The behaviour of these generators is better visualised when considering their action on strands. Specifically, if n strands are placed in a canonical order, then the element b_i describes the effect of exchanging the strands i and $i + 1$ in a clockwise fashion, as shown in Figure 8.8. Their inverses are given in terms of counterclockwise exchanges. All possible braidings of strands can be written as a combination of the b_i's and their inverses. Consider, for example, the product of two braiding elements b_1 and b_2^2. This product can be represented by placing the diagrams of the corresponding generators on top of each other in a time-ordered fashion and glueing the aligned strands together, as shown in Figure 8.9. A general braid is obtained when multiplying the generators b_i together in arbitrary integer powers. This creates braidwords of the form

$$B = b_{i_1}^{k_1} b_{i_2}^{k_2} \ldots b_{i_m}^{k_m}, \tag{8.9}$$

where the powers of b's are positive or negative integers and the subscripts allow the b's to order in any desired way.

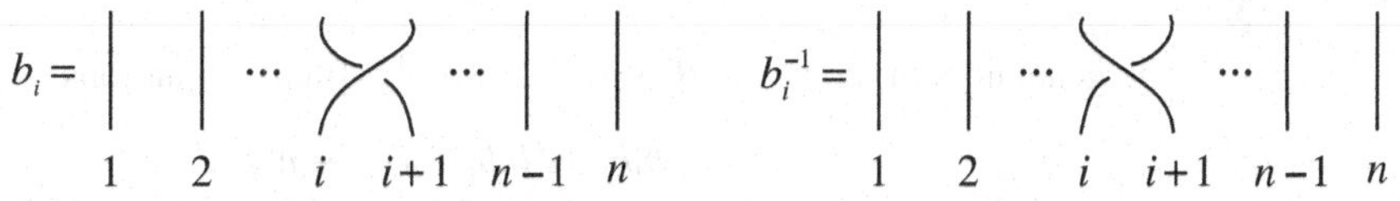

Fig. 8.8   The braiding generators, b_i, and their inverses, b_i^{-1}, for $i = 1, ..., n - 1$, act on n strands. (a) The effect of b_i is to exchange the strands i and $i + 1$ in a clockwise way, where time runs from top to bottom. (b) The braid b_i^{-1} exchanges the strands i and $i + 1$ in a counterclockwise way.

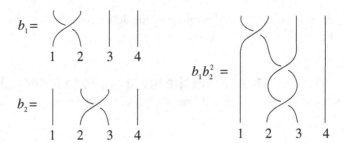

Fig. 8.9 The interpretation of braid generators as braided strands. The product of braiding generators is diagrammatically represented by placing the corresponding diagrams on top of each other. Here the product $b_1b_2^2$ is constructed out of the diagrams of b_1 and b_2. Concatenating such diagrams gives rise to a general braid of strands.

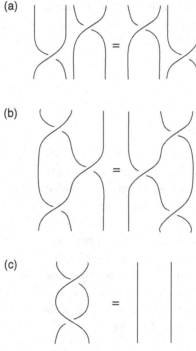

Fig. 8.10 Schematic representation of the Yang–Baxter equations. (a) Exchanging the order of two braids does not have an effect if they are sufficiently far apart, i.e., $b_ib_j = b_jb_i$ when $|i - j| \geq 2$. (b) Two braidings are equivalent under simple continuous deformations of the strands, $b_ib_{i+1}b_i = b_{i+1}b_ib_{i+1}$. (c) Undoing a braid gives the identity $b_ib_i^{-1} = e$.

The generators of the group B_n satisfy the Yang–Baxter equations

$$b_ib_j = b_jb_i, \text{for } |i - j| \geq 2, \tag{8.10}$$

$$b_ib_{i+1}b_i = b_{i+1}b_ib_{i+1}, \tag{8.11}$$

$$b_ib_i^{-1} = b_i^{-1}b_i = e, \tag{8.12}$$

where e is the identity element of the group. These relations have a simple diagrammatic interpretation, which can be found in Figure 8.10. Even though we can represent the braids with diagrams we should not forget that we are actually interested in their matrix representation.

The Yang–Baxter equations admit as solutions an infinite number of matrix representations of braids. They describe the effect braiding of anyons can have on their quantum states. The simplest such statistical evolution is the one-dimensional representation. Its elements $b_i = e^{i\theta}$, where $\theta \in [0, 2\pi)$, correspond to Abelian anyons. Non-Abelian anyons correspond to higher-dimensional irreducible representation of the braid group. They can produce non-trivial unitary rotations of the quantum states of anyons.

8.2.2 The Temperley–Lieb algebra

The braid group is a mathematical representation of braided strands. To derive topological invariant quantities related to the links of these strands, we need to decompose braided strands to simpler geometrical structures. Mathematically, this can be achieved with the help of the Temperley–Lieb algebra, which is diagrammatically equivalent to substituting the crossings of the strand braids with avoided crossings.

The Temperley–Lieb algebra $TL_n(d)$ is generated by E_i with $i = 1, \ldots, n-1$ that satisfy

$$E_i E_j = E_j E_i, \text{ for } |i - j| \geq 2, \tag{8.13}$$

$$E_i E_{i\pm1} E_i = E_i, \tag{8.14}$$

$$E_i^2 = d E_i, \tag{8.15}$$

where d is a real number. Apart from their matrix representation, these elements can be geometrically interpreted in terms of the Kauffman n-diagram, which is shown in Figure 8.11. A general Kauffman n-diagram has n canonically ordered points at the top and at the bottom of a rectangle. These points are connected to each other with strands. In particular, these strands do not cross each other and have no loops. Figure 8.12 illustrates the properties (8.13), (8.14) and (8.15) in terms of Kauffman n-diagrams.

The purpose of introducing the Temperley–Lieb algebra is to establish a useful unitary representation of the braid group. Let us define the particular representation ρ_A as

$$\rho_A(b_i) = A E_i + A^{-1} \mathbb{1} \text{ and } \rho_A(b_i^{-1}) = A^{-1} E_i + A \mathbb{1}, \tag{8.16}$$

where $\mathbb{1}$ is the identity and A is a newly introduced parameter. This is a natural representation of the braid group, if A satisfies $d = -A^2 - A^{-2}$. To see this we need to verify that the representation group elements, $\rho_A(b_i)$, satisfy the braid group relations (8.10), (8.11) and

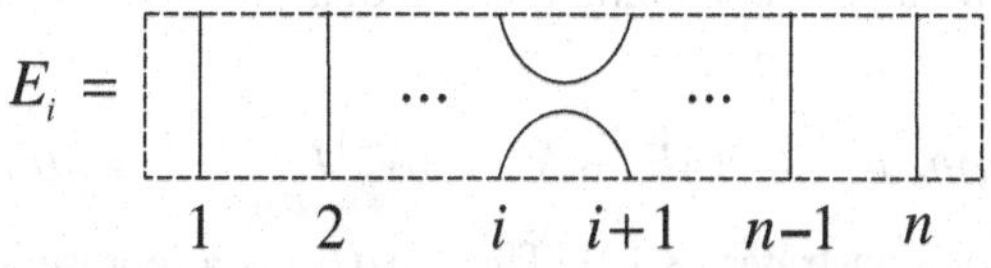

$$E_i = $$ 1   2 i $i+1$ $n-1$ n

Fig. 8.11   The Temperley–Lieb algebra element E_i for $i = 1, \ldots, n - 1$ can be represented by a Kauffman n-diagram. This is a rectangle with n points at the top and the bottom connected with straight lines, apart from the i and $i + 1$ elements, that are connected at each side with each other.

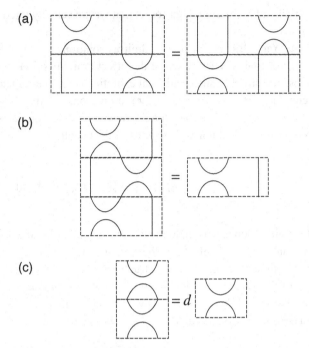

(a)

(b)

(c)

Fig. 8.12 Diagrammatic representation of the Temperley–Lieb algebra properties. (a) $E_i E_j = E_j E_i$ for $|i - j| \geq 2$, (b) $E_i E_{i\pm1} E_i = E_i$ and (c) $E_i^2 = dE_i$.

(8.12) provided the E_i's satisfy (8.13), (8.14) and (8.15). Let us demonstrate this equivalence. It is easy to see that the representations of b_i and b_j commute as E_i and E_j commute, for $|i - j| \geq 2$. Indeed, for $|i - j| \geq 2$ we have

$$\rho_A(b_i)\rho_A(b_j) = (AE_i + A^{-1}\mathbb{1})(AE_j + A^{-1}\mathbb{1}) = (AE_j + A^{-1}\mathbb{1})(AE_i + A^{-1}\mathbb{1})$$
$$= \rho_A(b_j)\rho_A(b_i), \tag{8.17}$$

which demonstrates (8.10). Subsequently, we want to show that $\rho_A(b_i)\rho_A(b_{i+1})\rho_A(b_i) = \rho_A(b_{i+1})\rho_A(b_i)\rho_A(b_{i+1})$. For that we have

$$\rho_A(b_i)\rho_A(b_{i+1})\rho_A(b_i)$$
$$= A^3 E_i E_{i+1} E_i + AE_{i+1}E_i + AE_i^2 + A^{-1}E_i + AE_iE_{i+1} + A^{-1}E_{i+1} + A^{-1}E_i + A^3$$
$$= (A^3 + Ad + A^{-1})E_i + 2AE_iE_{i+1} + A^{-1}(E_i + E_{i+1}) + A^3$$
$$= \rho_A(b_i)\rho_A(b_{i+1})\rho_A(b_i), \tag{8.18}$$

where we used the defining relation of d that gives $A^3 + Ad + A^{-1} = 0$. This demonstrates (8.11). Finally, we have

$$\rho_A(b_i)\rho_A(b_i^{-1}) = (AE_i + A^{-1}\mathbb{1})(A^{-1}E_i + A\mathbb{1}) = dE_i + A^2 E_i + A^{-2}E_i + \mathbb{1} = \mathbb{1}, \quad (8.19)$$

which demonstrates (8.12). Thus, $\rho_A(b_i)$ is a representation of the braid group generators B_n parameterised by A. To ensure that the representation $\rho_A(b_i)$ is unitary we need to take $|A| = 1$ and E_i Hermitian for all i. Indeed, then we have

$$\rho_A(b_i)\rho_A(b_i)^\dagger = (AE_i + A^{-1}\mathbb{1})(A^*E_i^\dagger + (A^{-1})^*\mathbb{1}) = E_i^2 + (A^2 + A^{-2})E_i + \mathbb{1} = \mathbb{1}. \quad (8.20)$$

The representation (8.16) can be viewed as the matrix equivalent of the Skein relations presented in Figure 8.3. The introduction of the Temperley–Lieb algebra decomposes a braidword to a sum of products of E_i elements. The corresponding Kauffman n-diagram of each term in the sum comprises a set of strands that do not cross each other. This sum is very similar to the state sum (8.3), though we still deal with Temperley–Lieb algebra elements and not with polynomials. To actually obtain a number we need to introduce an appropriate trace.

8.2.3 Markov trace and Jones polynomials

The concept of a trace is well defined for the case of a square matrix, as the sum of all its diagonal elements. We can generalise the trace as a mapping from braids or Temperley–Lieb elements to numbers. Then the trace of the braids $\rho_A(b)$, parameterised by A, is, in general, a polynomial in A. In addition, we can give a geometrical interpretation to this trace that establishes the relation between open-ended strands and links. A version of this tracing procedure, called the Markov trace, consists of connecting the opposite endpoints of a braidword B together, as shown in Figure 8.13(a). We denote the resulting link

$$L = (B)^{\text{Markov}}. \quad (8.21)$$

In conclusion, a braidword with a trace gives a link. Maybe it is less expected that the converse is also true. From Alexander's theorem (1923) we know that every link can be obtained from a braidword with a trace.

We would like to evaluate the Markov trace of braidwords with braid group elements represented by (8.16). For that we need the Markov trace of a product of elements of the Temperley–Lieb algebra, $TL_n(d)$, that comprise a Kauffman n-diagram. In the following, we call such a product, as well as its corresponding diagram, K. The Markov trace on K can be performed again by connecting its opposite endpoints together. This gives rise to a set of disjoint loops, as shown in Figure 8.13(b). Let a be the number of such loops. Then we can define the trace of K to be

$$\text{tr}(K) = d^{a-n}, \quad (8.22)$$

(a)   (b)

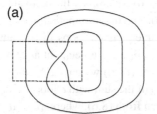 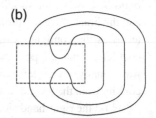

Fig. 8.13 Diagrammatically, the Markov trace corresponds to connecting together the opposite ends of the strands. (a) The Markov trace of a braid element. (b) The Markov trace of a Temperley–Lieb element.

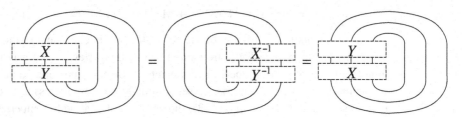

Fig. 8.14 The trace of two general Temperley–Lieb elements is commutative, i.e., $\text{tr}(XY) = \text{tr}(YX)$. Indeed, by continuous deformations of the strands we can move X and Y on the other side and then rotate the whole diagram by π to bring it back to the original orientation. These processes exchange the order of X and Y.

where n is the number of points at a horizontal side of the Kauffman diagram. From (8.22) we can derive the following properties of the Markov trace:

$$\text{tr}(\mathbb{1}) = 1, \tag{8.23}$$

$$\text{tr}(XY) = \text{tr}(YX), \quad \text{for any } X, Y \in TL_n(d), \tag{8.24}$$

$$\text{tr}(XE_{n-1}) = \frac{1}{d}\text{tr}(X), \quad \text{for any } X \in TL_{n-1}(d). \tag{8.25}$$

These properties follow directly from the definition (8.22). Indeed, the trace of the Kauffman diagram of $\mathbb{1}$ gives $a = n$ loops, so (8.23) follows. The commutation process of two elements X and Y is shown diagrammatically in Figure 8.14, thus demonstrating (8.24). Finally, if a single E_{n-1} element is at the end of a Kauffman diagram then its removal produces an extra loop component and a d contribution needs to be removed. This can be verified with the example of Figure 8.13(b). There, if we replace E_2 with the identity then we need to introduce a factor $1/d$ in (8.25) to compensate for the change in the numbers of loops. Moreover, it is possible to show that the Markov trace satisfying properties (8.23), (8.24) and (8.25) is uniquely defined (Aharonov *et al.*, 2009).

The Markov trace acts linearly, i.e., $\text{tr}(K + K') = \text{tr}(K) + \text{tr}(K')$. So we can consider the trace of braids in the representation $\rho_A(B)$ of (8.16) as actually a trace acting on a sum of general Temperley–Lieb elements. Moreover, one can show that this trace relates directly to the state sum (8.6) as follows (Aharonov *et al.*, 2009):

$$d^{n-1}\text{tr}(\rho_A(B)) = \langle (B)^{\text{Markov}} \rangle(A). \tag{8.26}$$

The link $L = (B)^{\text{Markov}}$ is defined as the diagrammatic representation of the Markov traced braid B. To show the validity of (8.26) we employ the equivalence between Skein relations and the braid representation (8.16). The diagrammatic trace of $\rho_A(B)$ has exactly the same terms as the configurations S of unentangled loops in the state sum (8.6). Also, all the coefficients that are powers of A are in exact agreement as they are introduced in the same way. Finally, on the left-hand side of (8.26) we have a factor of d^{n-1} and a factor of d^{a-n} coming from the trace of each term of the sum with a loops. This exactly matches the factor $d^{|S|-1}$ on the right-hand side as seen from (8.6), where $|S| = a$. So (8.26) holds identically.

From the state sum we can write the Jones polynomial directly in the following way:

$$V_L(A) = (-A)^{3w(L)}d^{n-1}\text{tr}(\rho_A(B)). \tag{8.27}$$

Thus, we have an explicit form for the Jones polynomials in terms of the braid representations $\rho_A(B)$. This was our initial goal. The importance of this relation is that it gives the means to calculate the Jones polynomials in terms of braiding operations. When a particular anyonic model is considered, with statistical evolutions given by $\rho_A(B)$, then $V_L(A)$ with $L = (B)^{\text{Markov}}$ can be calculated from the evolution operator of anyons. This is analysed in detail in the next section.

8.3   Analogue quantum computation of Jones polynomials

Before we demonstrate how to obtain Jones polynomials from anyonic quantum evolutions, let us first remark on their importance. The practical interest in evaluating Jones polynomials is based on the following fact. A link L gives rise to a certain Jones polynomial, $V_L(A)$, parameterised by A. This polynomial is constructed so that another link L' that can be continuously deformed to L gives exactly the same Jones polynomial, $V_{L'}(A) = V_L(A)$. Importantly, when two links have different values of Jones polynomials for some A, they are inequivalent. The reason for this is that if they were equivalent, they would necessarily give the same Jones polynomial. However, equality in the Jones polynomials of two links does not imply that the links are equivalent. These relations are summarised in Table 8.1.

We do not know of any topological invariant that uniquely characterises links and distinguishes all the non-equivalent ones. Finding link invariants with this property is one of the main goals in the mathematical field of topology. Still, being able to efficiently evaluate the Jones polynomials for some A gives the means to distinguish inequivalent links. This can be very useful in various areas of statistical physics, applied technologies and medicine, where complex strand-like structures emerge. Indeed, the braid configuration of extended objects determines their properties in many cases.

Even the approximate computation of Jones polynomials is a **BQP**-complete problem as it requires all the power quantum mechanics can offer. We know that the exact evaluation of Jones polynomials would take exponential classical resources (Jaeger *et al.*, 1990). Note that no closed form exists for this task apart from a handful of values of A. Nevertheless, it is possible to extract information about the value of the Jones polynomial by simply braiding anyons together. Such a quantum simulation with anyons was proposed by Freedman *et al.*, (2003b). It takes advantage of the quantum properties of anyons in order to

Table 8.1 Relation between links and Jones polynomials		
Links		Jones polynomials
$L = L'$	$\Longrightarrow$	$V_L(A) = V_{L'}(A)$
$L \neq L'$	$\Longleftarrow$	$V_L(A) \neq V_{L'}(A)$

Table 8.2 Quantum simulation with anyons		
Quantum simulation		**Anyonic evolution**
State initialisation	$\longleftrightarrow$	Pair creation of anyons
Quantum evolution	$\longleftrightarrow$	Anyon braiding
Readout	$\longleftrightarrow$	Anyonic fusion

efficiently evaluate the Jones polynomials (Freedman *et al.*, 2002c). The translation of such an anyonic evolution to a quantum algorithm was performed by Aharonov *et al.* (2009).

We now describe the main principles in the analogue quantum computation of Jones polynomials with anyons, which is outlined in Table 8.2. We initially consider an anyonic model with statistics that corresponds to a particular representation of the braiding group, $\rho_A(B)$, given in (8.16). The parameter A provides the particular statistical behaviour of the anyons. The expectation values of their braiding evolutions can then be related to the Jones polynomials in the following way. Consider an anyonic evolution, such as the one given in Figure 8.15. From the vacuum, create n anyons in pairs. This initial state, denoted by $| \psi \rangle$, signifies that each anyonic pair is in the vacuum fusion channel. The state that corresponds to fusing these anyons to the vacuum is denoted by $\langle \psi |$. The probability that they fuse back to the vacuum in the same pairwise order is given by $\langle \psi | \psi \rangle = 1$, as expected. Suppose we now perform an arbitrary braiding among half of the anyons, as in Figure 8.15, described by the unitary evolution $B(A)$. We then pairwise fuse them with the same ordering as the pair creation. The probability of obtaining the vacuum state at the end is then given by

$$\langle \psi \, | \, B(A) \, | \, \psi \rangle = \mathrm{tr}(\rho_A(B)). \tag{8.28}$$

Relation (8.28) expresses the amplitude of the braiding evolution in terms of a quantum mechanical expectation value and in terms of the trace of the braid representation $\rho_A(B)$. Together with (8.27) we can hence express the Jones polynomials in terms of probability outcomes

$$V_L(A) = (-A)^{3w(L)} d^{n-1} \, \langle \psi \, | \, B(A) \, | \, \psi \rangle . \tag{8.29}$$

The probabilities $\langle \psi \, | \, B(A) \, | \, \psi \rangle$ can be obtained experimentally from an anyonic system with an accuracy that improves by repeating the same experiment several times. To calculate this expression, we need to know the writhe of the link $L = (B)^{\mathrm{Markov}}$. From (8.8) we see that the writhe is an additive quantity that can be calculated with polynomial resources (Aharonov *et al.*, 2009).

In essence, what we actually compute is the trace of the braiding matrices without the need to evaluate each diagonal element of the braidword. The described anyonic evolution is designed exactly for this purpose. As an alternative to the anyonic quantum simulation, good quantum algorithms exist also for computing traces of unitary matrices. The algorithmic approach is based on the Hadamard test (Aharonov *et al.*, 2009). We can begin with

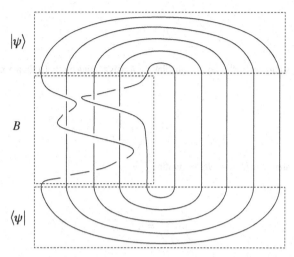

An anyonic quantum evolution that serves as an analogue computation of Jones polynomials, where time flows downwards. Five pairs of anyons are created from the vacuum. This initial state is denoted by $|\psi\rangle$. Half of the anyons are braided with each other to produce the braidword B. Subsequently, the anyons are fused back together. The state of the anyons that corresponds to the vacuum fusion outcome is the same state $\langle\psi|$. The probability of this evolution is the state sum that directly gives the desired Jones polynomial.

a completely mixed state of n register qubits and one work qubit w prepared in the pure state

$$| \psi_w \rangle = \frac{1}{\sqrt{2}} \left(| 0 \rangle_w + | 1 \rangle_w \right).$$ (8.30)

We then apply a sequence of controlled unitaries

$$U = \prod_{j=1}^{m} | 1 \rangle_w \langle 1 | \otimes \rho_A(b_j).$$ (8.31)

Measuring the work qubit in the x and y directions of the Bloch sphere finally gives the real and imaginary parts of the normalised trace $\mathrm{tr}(\rho_A(B))/2^n$, respectively (see Exercise 8.2).

8.4 Example I: Kauffman bracket of simple links

To familiarise ourselves with the Kauffman bracket or state sum we now evaluate it for some simple links. Our first example, L_1, is the 'figure of eight' in Figure 8.16. Its state sum is given by

$$\langle L_1 \rangle = (-A)^{-3}$$ (8.32)

as it involves a single twist. Similarly for the inverted twisting, L_2, of Figure 8.17 we have the value

$$\langle L_2 \rangle = (-A)^3.$$ (8.33)

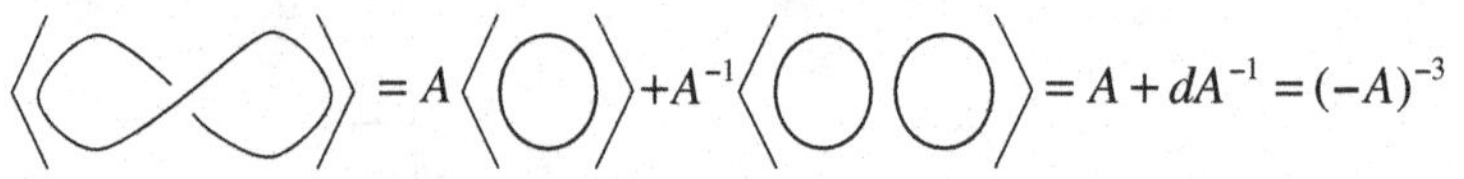

$$\left\langle \infty \right\rangle = A \left\langle \bigcirc \right\rangle + A^{-1} \left\langle \bigcirc\bigcirc \right\rangle = A + dA^{-1} = (-A)^{-3}$$

Fig. 8.16   The state sum for the 'figure eight' link, L_1. As it is a single twist of a simple loop it gives $\langle L_1 \rangle = (-A)^{-3}$.

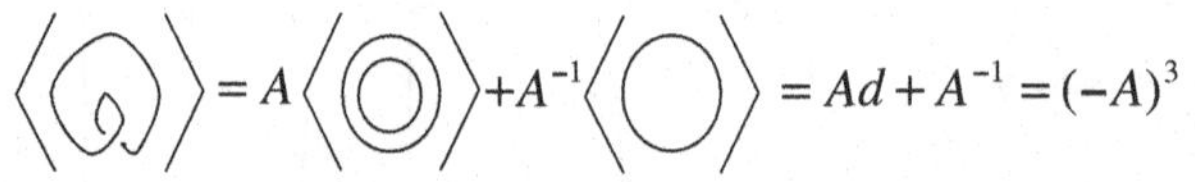

$$\left\langle \bigcirc \right\rangle = A \left\langle \bigcirc \right\rangle + A^{-1} \left\langle \bigcirc \right\rangle = Ad + A^{-1} = (-A)^3$$

Fig. 8.17   The state sum for the inverted twisting, L_2 is $\langle L_2 \rangle = (-A)^3$.

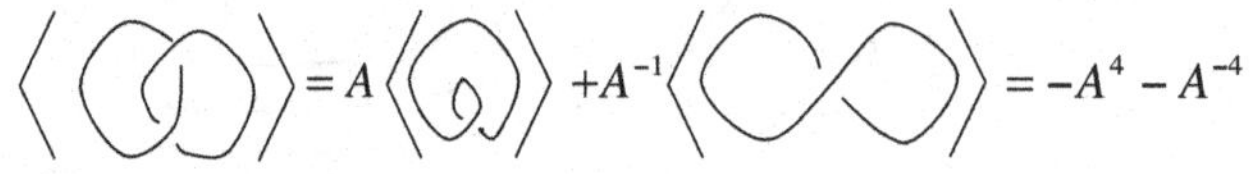

$$\left\langle \bigcirc \right\rangle = A \left\langle \bigcirc \right\rangle + A^{-1} \left\langle \infty \right\rangle = -A^4 - A^{-4}$$

Fig. 8.18   The simple non-trivial link, L_3, with two components has $\langle L_3 \rangle = -A^4 - A^{-4}$.

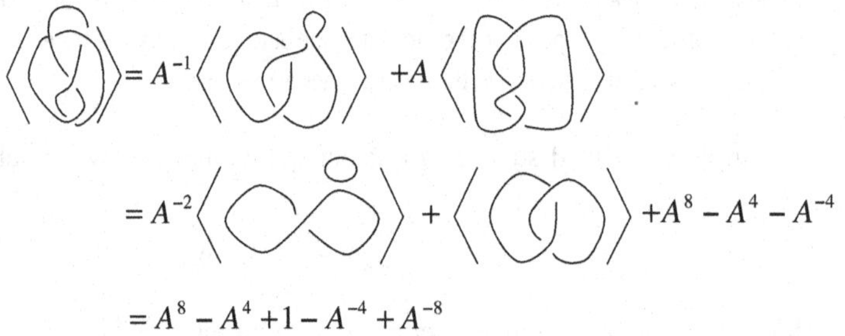

$$\left\langle \bigcirc \right\rangle = A^{-1} \left\langle \bigcirc \right\rangle + A \left\langle \bigcirc \right\rangle$$

$$= A^{-2} \left\langle \bigcirc \right\rangle + \left\langle \bigcirc \right\rangle + A^8 - A^4 - A^{-4}$$

$$= A^8 - A^4 + 1 - A^{-4} + A^{-8}$$

Fig. 8.19   A non-trivial single component link, L_4 with $\langle L_4 \rangle = A^8 - A^4 + 1 - A^{-4} + A^{-8}$.

We now consider the link with two components, L_3, shown in Figure 8.18. By employing the state sums of the previous examples we easily obtain

$$\langle L_3 \rangle = -A^4 - A^{-4}. \tag{8.34}$$

Finally, we evaluate the link, L_4, in Figure 8.19. Its state sum is given by

$$\langle L_4 \rangle = A^8 - A^4 + 1 - A^{-4} + A^{-8}. \tag{8.35}$$

To evaluate the Jones polynomials of these links we need to apply relation (8.7) that states $V_L(A) = (-A)^{3w(L)}\langle L \rangle$. The first two links, L_1 and L_2, have $w = 1$ and $w = -1$ respectively, so $V_{L_1}(A) = 1$ and $V_{L_2}(A) = 1$. Hence they have the same Jones polynomials for any A. This is to be expected as they are both isomorphically equivalent to a simple loop. On the other hand, $w(L_3) = 2$ and $w(L_4) = 0$.

8.5   Example II: Jones polynomials from Chern—Simons theories

Here we investigate how the Jones polynomials can be derived from the SU(2) Chern–Simons theories that we studied in Chapter 7. The specific form of the Jones polynomials was determined by introducing the Skein relations, in Figure 8.7. Here, we demonstrate that the expectation values of Wilson loops, $\langle W(L) \rangle$, in the SU(2) Chern–Simons theories can be decomposed in the same way as the state sums under Skein relations. This decomposition is compatible with the Reidemeister moves II and III. The invariance of this expectation value under continuous deformations of the loop L means that $\langle W(L) \rangle$ is invariant under twists of the loop as well. This property is the Reidemeister move I that finally identifies $\langle W(L) \rangle$ with the Jones polynomials.

Let us see in detail how the SU(2) Chern–Simons theory is compatible with the Skein relations. Consider the expectation value $\langle W(L) \rangle$ of a link L in space $M = S^3$. We take all link components to be in the two-dimensional fundamental representation of SU(2). A useful bipartition of the link L is given in Figure 8.20(a). There, one part, L_R, includes a single crossing of two strands and the other part, L_L, includes the rest of the link. The corresponding spaces are denoted M_R and M_L, respectively. Substituting the crossing in M_R with any of the two shapes in M'_R or M''_R undoes the braiding between the two relevant strands and gives a simpler link. In terms of the expectation value $\langle W(L) \rangle$ this substitution is motivated in the following way. Consider the individual parts M_L and M_R. Each one supports a

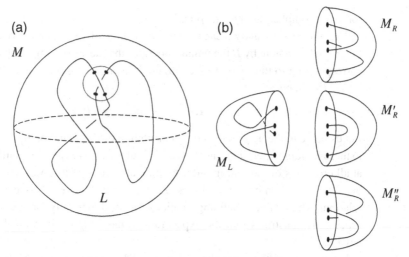

Fig. 8.20   (a) The link L can be split into two parts by separating space M into M_L and M_R. The M_R part is the small dotted sphere that includes a single crossing of the link. (b) If we substitute the space M_R on the right with M'_R or M''_R then the linked components of L separate out, i.e., the composite of the space M_L on the left with M'_R or M''_R gives a trivial one- or two-component link.

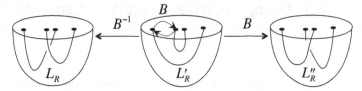

Consider the parts of the link L_R, L_R' and L_R'' that correspond to M_R, M_R' and M_R'', respectively. The clockwise exchange of the first two endpoints of L_R' gives rise to L_R''. The counterclockwise exchange of these two points gives rise to L_R.

$$B = \bigtimes \qquad B^{-1} = \bigtimes \qquad \mathbb{1} = \Big| \ \Big|$$

The correspondence between link configurations and braid operators. In particular, B gives a clockwise exchange of two points, B^{-1} is counterclockwise and $\mathbb{1}$ keeps them still.

two-dimensional Hilbert space $\mathcal{H}_R$, as they correspond to the fusion of the four points that are given by the intersections of the link and the dotted sphere, shown in Figure 8.20(a). As these points are all described by the fundamental representation of SU(2), as in (7.56), they have only two possible fusion outcomes. So the Hilbert space $\mathcal{H}_R$ is two-dimensional. Let us denote the vectors that correspond to M_R, M_R' and M_R'' by ψ, ψ' and ψ'', respectively. As all these three vectors belong to the same two-dimensional Hilbert space they are linearly dependent, i.e., we can write

$$\alpha \psi + \beta \psi' + \gamma \psi'' = 0, \tag{8.36}$$

for some complex numbers α, β and γ.

The link parts L_R and L_R'' can be obtained from L_R' by exchanging two points, as shown in Figure 8.21. Denote by B the unitary operator that describes this exchange. Then the states corresponding to the L_R and L_R'' configurations are given by $\psi = B^{-1}\psi'$ and $\psi'' = B\psi'$. Hence (8.36) becomes

$$\alpha B^{-1}\psi' + \beta \psi' + \gamma B\psi' = 0. \tag{8.37}$$

Let us move to the diagrammatic representation of this equation by assigning parts of links to these braiding operations, as shown in Figure 8.22. We can use relation (8.37) recursively at all crossings of the link in order to undo all braids of its strands and reduce it to a sum of completely non-entangled loops. Assume now that (8.37) is equivalent to the Skein relations which are the defining relations for the Jones polynomials. Then, this reduction process demonstrates that the expectation value $\langle W(L) \rangle$ gives the Jones polynomial for any link.

We now show the equivalence between (8.37) and the Skein relations. As $\langle W(L) \rangle$ is invariant under continuous deformations of the link L, the decomposition (8.37) has to be compatible with the Reidemeister moves II and III. Following a similar process to the one we used to determine the coefficients of the Skein relations in Section 8.1.2, we obtain

$$\alpha = A, \ \beta = -(A^2 - A^{-2}) \text{ and } \gamma = -A^{-1} \tag{8.38}$$

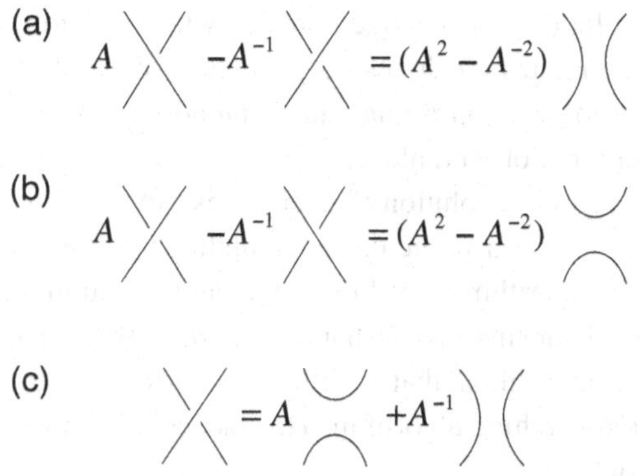

(a)
$$A \;\times\; -A^{-1}\;\times\; = (A^2 - A^{-2}) \;)\,($$

(b)
$$A \;\times\; -A^{-1}\;\times\; = (A^2 - A^{-2}) \;\cup \atop \cap$$

(c)
$$\times \; = A \;{\cup \atop \cap}\; + A^{-1}\;)\,($$

Fig. 8.23   (a) Diagrammatic representation of (8.37) for $\alpha = A$, $\beta = -(A^2 - A^{-2})$ and $\gamma = -A^{-1}$. Rotating the (a) graph clockwise by $\pi/2$ gives (b). A linear combination of (a) and (b) gives (c), which is the familiar Skein relation, given in Figure 8.7.

(see Exercise 8.3). Now, let us consider the diagrammatic form of (8.37), which is parameterised by A, as shown in Figure 8.23(a). Rotating the diagram in Figure 8.23(a) clockwise by $\pi/2$ gives Figure 8.23(b). Linear combinations of these two equations give Figure 8.23(c). This is indeed the Skein relation presented in Figure 8.7. For consistency of (8.37) with respect to Reidemeister move II we also have $\langle W(L \cup \bigcirc)\rangle = -(A^2 + A^{-2})\langle W(L)\rangle$, where L is an arbitrary link (see Exercise 8.3). Note that the brackets now denote expectation values rather than the state sums we saw in (8.1). Hence, the expectation value $\langle W(L)\rangle$ of the SU(2) Chern–Simons theory gives the Jones polynomial. It has been shown by explicit evaluation of the expectation value that it is $A = ie^{i\frac{\pi}{2(k+2)}}$ for the SU(2) level-k theory (Witten, 1989).

Summary

The quantum simulation of Jones polynomials manifests a beautiful connection between mathematics, physics and information science. Jones polynomials are link invariants that remain unchanged when we apply continuous link deformations. Such deformations might produce link shapes that seem completely inequivalent to the original one. Hence, being able to efficiently compute the Jones polynomials can be very useful in many research disciplines that investigate the properties of extended objects.

To present the Jones polynomials we used two approaches. We first assigned numbers to geometrical characteristics of the links in such a way that the final number remains invariant under continuous transformations of the link strands. This approach demonstrated that the Jones polynomials are actually link invariants. Then we derived these polynomials with an algebraic approach. We used the braid group to assign matrix representations to the braided strands of a link. The resulting braiding matrix actually describes the quantum state evolution of exchanged anyons whose worldlines span the strands of the link. This approach relates the Jones polynomials with a physical observable. From this point it is possible to envision a quantum simulation with anyons that can extract the expectation value of these observables.

The anyonic evolution that gives rise to the Jones polynomials is a quantum simulation that solves a particular computationally hard problem (Freedman *et al.*, 2003b). A quantum algorithm, based on qubits and quantum gates instead of anyons, has also been developed for this task (Aharonov *et al.*, 2009). This algorithm provides a new class of quantum algorithms that is different in structure from the Shor's factoring algorithm and Grover's searching algorithm. This fact could be proven useful in addressing new algorithmic problems.

Exercises

8.1 Demonstrate that the Kauffman bracket of the trefoil is given by $-A^5 - A^{-3} + A^{-7}$. What is the corresponding Jones polynomial?

8.2 Show how the Hadamard test presented in (8.30) and (8.31) works for the case of a two-dimensional unitary matrix.

8.3 Demonstrate that requiring (8.37) to be compatible with the Reidemeister moves II and III gives (8.38) and that $\langle W(L \cup \bigcirc) \rangle = -(A^2 + A^{-2})\langle W(L) \rangle$, where L is an arbitrary link. [*Hint*: It might be easier to give (8.37) a form similar to the Skein relations.]

9

Topological entanglement entropy

To perform topological quantum computation we first need to experimentally realise anyons in a topological system. These systems are characterised by intriguing non-local quantum correlations that give rise to the anyonic statistics. What are the diagnostic tools we have to identify if a given system is indeed topological or not? Different phases of matter are characterised by their symmetries. This information is captured by order parameters. Usually, order parameters are defined in terms of local operators that can be measured in the laboratory. For example, the magnetisation of a spin system is given as the expectation value of a single spin with respect to the ground state. Such local properties can describe fascinating physical phenomena efficiently, such as ferromagnetism, and can pinpoint quantum phase transitions.

But what about topological systems? Experimentally, we usually identify the topological character of systems, such as the fractional quantum Hall liquids, by probing the anyonic properties of their excitations (Miller *et al.*, 2007). However, topological order should be a characteristic of the ground state (Thouless *et al.*, 1982; Wen, 1995). The natural question arises: is it possible to identify a property of the ground state of a system that implies anyonic excitations? The theoretical background that made it possible to answer this question came from entropic considerations of simple topological models. Hamma *et al.* (2005) studied the entanglement entropy of the toric code ground state and noticed an unusual behaviour. Even though the ground state has a short correlation length due to the energy gap above it, long-range correlations of topological origin are also present. Consequently, Kitaev and Preskill (2006) and simultaneously Levin and Wen (2006) introduced the concept of topological entanglement entropy or just topological entropy. This quantity successfully identifies long-range correlations of topological origin.

As topological entropy can identify if a given system is topologically ordered or not, it can be used to study the transitions between topologically ordered and non-ordered phases of matter. It also offers a quantitative criterion to study how temperature or external perturbations affect topological order. However, we should bear in mind that the resilience of topological quantum computation is not necessarily identical with the resilience of topological order. A computation allows for active manipulations that could potentially neutralise the effect of errors.

In the following we give an overview of the entanglement entropy of two-dimensional systems. Its striking characteristic is that topological order naturally arises as a particular case of the few possible entropic behaviours. Next we present the definitions of topological order given by Kitaev and Preskill (2006) as well as Levin and Wen (2006). Although these approaches are equivalent, each one offers a different perspective on topological entropy.

Finally, we derive the connection between topological order and entanglement entropy for the case of quantum double models (Iblisdir *et al.*, 2009).

9.1 Entanglement entropy and topological order

Before we introduce the concept of topological entropy, we define the entanglement entropy and we examine some of its generic properties. Entanglement entropy is defined in terms of the von Neumann entropy

$$S(\rho) = -\mathrm{tr}(\rho \log \rho) \tag{9.1}$$

of a density matrix ρ in the following way. Consider a two-dimensional interacting many-particle system prepared in its ground state and assume there is a non-zero energy gap above it. We take the system to be in a pure state, i.e., its temperature is zero and it is isolated from its environment. Due to the energy gap the ground state $| \Psi \rangle$ of the system has a finite correlation length, ξ. This length parameterises the exponential decay of two-point correlations of the ground state in terms of local observables. To be more explicit, we now consider an operator $O(\mathbf{r})$ defined at the neighbourhood of a point $\mathbf{r}$ of the system. For a gapped system we expect that

$$\langle O(\mathbf{r}_1)O(\mathbf{r}_2)\rangle - \langle O(\mathbf{r}_1)\rangle\langle O(\mathbf{r}_2)\rangle \sim \exp\left(-\frac{|\mathbf{r}_1 - \mathbf{r}_2|}{\xi}\right), \tag{9.2}$$

where the expectation values are with respect to the ground state $| \Psi \rangle$. This relation signifies that correlations between any two points are exponentially suppressed with respect to their relative distance.

We now consider the bipartition of a system in A and its complement B, separated by the boundary ∂A, as shown in Figure 9.1. Suppose the density matrix of the system is given by

$$\rho = | \Psi \rangle \langle \Psi |. \tag{9.3}$$

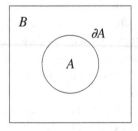

Fig. 9.1   A two-dimensional quantum system prepared in its ground state is partitioned into two regions, A and B, via the boundary ∂A. The entanglement entropy of the reduced part A is given, up to small corrections, by $S_A = \alpha|\partial A| - \gamma$, where α and γ are system-dependent constants and $|\partial A|$ is the length of the boundary.

Denote by ρ_A the reduced density matrix of A obtained by tracing the part of the system in the complement B, i.e.,

$$\rho_A = \text{tr}_B \rho. \qquad (9.4)$$

The entanglement entropy of A is then defined as the von Neumann entropy of ρ_A, i.e.,

$$S_A = -\text{tr}\,(\rho_A \log \rho_A)\,. \qquad (9.5)$$

The logarithm of ρ_A can be calculated from its eigenvalues λ_i by bringing the entropy finally into the form

$$S_A = -\sum_i \lambda_i \log \lambda_i. \qquad (9.6)$$

This entropy can be used to measure the correlations of the geometrical region A with respect to its environment B.

Assume now that the entanglement entropy can be written as

$$S_A = \alpha |\partial A| - \gamma + \epsilon(|\partial A|^{-\beta}), \qquad (9.7)$$

where α, γ are real numbers and $\beta > 0$. So $\epsilon(|\partial A|^{-\beta})$ tends to zero as the size of the boundary, $|\partial A|$, tends to infinity. By this limit we mean that all relevant scales in the shape of ∂A, such as its radius if it is a circle or its smallest side if it is a rectangle, are much larger than the correlation length, ξ, of the ground state. Relation (9.7) provides a generic description of the entropy. Let us justify this form for S_A. No 'volume' term, $|A|$, appears as our system is in a pure state. Indeed, if every particle of the system was entangled with the environment then it would individually contribute to the entropy and a total term proportional to $|A|$ would emerge.

Let us now motivate the 'area' term $|\partial A|$ that describes the behaviour of gapped systems. If $|\Psi\rangle$ was a tensor product state between the A and B parts, say $|\Psi\rangle = |\Psi_A\rangle \otimes |\Psi_B\rangle$, then $S_A = 0$. The entropy S_A only becomes non-zero if the two regions are non-trivially entangled. In other words, the entanglement entropy, $S(\rho_A)$, measures to what extent part A of the system is entangled with part B. As the correlation length of the system is finite, we cannot expect a particle deep in the A region to be entangled with a particle deep in B. Hence, the only contribution to the entanglement between A and B should come from the regions near the boundary, ∂A, as shown in Figure 9.2. Actually, we expect a strip at each side of ∂A, as wide as the correlation length, to be responsible for the main contribution to $S(\rho_A)$, which should be proportional to $|\partial A|$. The coefficient α in (9.7) is a non-universal quantity that depends in general on the small-scale properties of the ground state, such as the correlation length ξ.

Of particular interest to us is the possible presence of a constant term γ. As shown in Kitaev and Preskill (2006), Levin and Wen (2006), systems with a non-zero γ are topologically ordered. Indeed, γ is related to the total quantum dimension $\mathcal{D}$ of the model, in the following way:

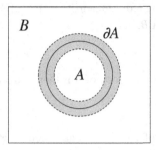

Fig. 9.2 As the system is gapped, the correlation length is finite. Hence, the correlations between regions A and B come predominantly from a strip around the boundary. The contributions to the entanglement entropy, S_A, mainly come from the correlations between the strip in A and the strip in B.

$$\gamma = \log \mathcal{D} = \log \sqrt{\sum_q d_q^2}. \qquad (9.8)$$

Here d_q is the quantum dimension associated with the anyon q and the summation runs through all the anyonic species of the model. In Section 9.3 we demonstrate this relation for the case of the quantum double models.

Let us consider the significance of γ in (9.7) in terms of the entropic properties of the system. In a sense, this constant term is also a boundary contribution. It measures the entropy associated with the non-local correlations between A and B, i.e., it characterises global features of the entanglement in the ground state. A non-zero value of γ shows that there is an additional order in the system that reduces the entropy by a constant amount. This order needs to be topological in nature as γ does not depend on the particular shape of the boundary.

From (9.8) we see that γ directly determines the quantum dimension $\mathcal{D}$. For example, when the system does not possess anyons, then $\mathcal{D} = 1$ as the quantum dimension gets contributions only from the 'vacuum'. As the vacuum corresponds to the state without anyons, we conclude that $\gamma = 0$ for systems that do not support anyons. Knowing the exact value of γ is not enough to uniquely determine a topological model as there can be several topological models with the same $\mathcal{D}$. Nevertheless, we now have a way to distinguish whether a system possesses topological order or not by determining whether $\gamma \neq 0$ or $\gamma = 0$, respectively.

9.2 Topological entropy and its properties

To quantify the non-local properties of entanglement that occur in topological systems we employ the von Neumann entropy of a reduced density matrix. Starting from (9.7), it is possible to distinguish the topological features from the non-topological ones. Indeed, from the entropy of particular geometric configurations we can isolate the constant term γ, as we shall see in the following.

9.2.1 Definition of topological entropy

We initially present the approach taken by Kitaev and Preskill (2006). Consider a gapped system defined on a closed surface Σ, e.g., the surface of a sphere. We partition Σ into four regions, A, B, C and D, as shown in Figure 9.3. Care is taken that the typical size of each partition is much larger than the correlation length of the ground state. Then the entanglement entropy is given by

$$S_X = \alpha|\partial X| - \gamma, \tag{9.9}$$

where X is region A, B, C or D. The topological part can be obtained from the following linear combination of entropies:

$$S_{\text{topo}} = S_A + S_B + S_C - S_{AB} - S_{AC} - S_{BC} + S_{ABC}, \tag{9.10}$$

where the entropy S_{AB} is evaluated for the composite region of A and B and so on. We call S_{topo} the topological entanglement entropy. By direct substitution of (9.7) we can show that

$$S_{\text{topo}} = -\gamma. \tag{9.11}$$

Indeed, the contributions in S_{topo} from all the boundary terms $|\partial X|$ are cancelled out. For example, the boundary between A and D in Figure 9.3 appears in S_A and S_{ABC} with positive sign, while it appears with negative sign in $-S_{AB}$ and $-S_{AC}$. On the other hand, the constant term appears three times with positive sign and four times with negative, giving eventually a net γ contribution. Four regions is the least number we can use to isolate the topological contribution (Iblisdir *et al.*, 2009).

Definition (9.10) provides a way to isolate the constant term γ by evaluating the entropies of different regions of a system. If S_{topo} is zero, then no topological order exists. If it is non-zero, then the system is necessarily topologically ordered, though loopholes for

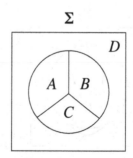

Fig. 9.3   The surface Σ partitioned into the regions A, B, C and D.

this test of topological order and their resolutions have been pointed out (Wootton, 2011). The order parameter γ has been employed as a defining relation for the study of topological order in a variety of systems, such as the fractional quantum Hall effect (Haque *et al.*, 2007) and Kitaev's honeycomb lattice model (Yao and Qi, 2010), as well as systems subject to temperature (Castelnovo and Chamon, 2007; Iblisdir *et al.*, 2009; Nussinov and Ortiz, 2008), generalisations to higher dimensions (Castelnovo and Chamon, 2008) and so on (Furukawa and Misguich, 2007; Hamma *et al.*, 2008; Papanikolaou *et al.*, 2007).

9.2.2 Properties of topological entropy

From the explicit form $\gamma = \log \mathcal{D} = \log \sqrt{\sum_q d_q^2}$ it is easy to see that the topological entropy depends only on the type of anyonic species that can be supported in the system. In particular, S_{topo} is independent of the shape and size of the regions A, B, C and D employed in definition (9.10) as long as their typical size is much larger than the correlation length of the ground state. Only the topological configuration of these regions, which is encoded in the neighbouring relation between them, is relevant. We call this property the topological invariance of S_{topo}.

Notice that the detailed characteristics of the Hamiltonian, that give rise to the topological model, do not enter the definition of S_{topo}. The only condition is that any relevant scale in the system, such as the size of the partitioning regions, has to be large with respect to the correlation length, ξ, of the ground state. The details of a model at small length scales are not relevant to its topological behaviour. This characteristic signifies the universality of topological entropy as it is able to describe different Hamiltonians that give rise to the same topological behaviour. In particular, smooth deformations of a Hamiltonian cannot change S_{topo} if they are not inducing a quantum phase transition. These properties of topological entropy are demonstrated in the following with simple considerations.

9.2.2.1 Topological invariance

To demonstrate topological invariance we need to show that deformations in the shape of the regions, shown in Figure 9.4, do not alter the value of S_{topo} (Kitaev and Preskill,

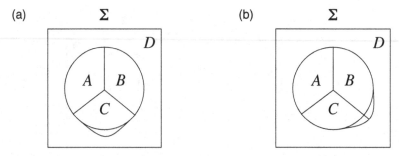

Fig. 9.4 (a) Deformation on the boundary between C and D. (b) Deformation in the position of the triple point between B, C and D. The topological entropy S_{topo} remains invariant under both (a) and (b) deformations.

2006). From the configuration of the partitions we see that there are only two possible distinct classes of boundary deformations. The first is changing one of the boundaries of the regions and the second is moving a point where three boundaries meet.

For concreteness, we deform the boundary between regions C and D, as shown in Figure 9.4(a), without affecting regions A or B. This boundary deformation changes the topological entropy by the amount

$$\Delta S_{\text{topo}} = (\Delta S_{ABC} - \Delta S_{BC}) - (\Delta S_{AC} - \Delta S_C). \tag{9.12}$$

If the above deformation is far enough from the other boundaries, the entropies S_A, S_B and S_{AB} change only by a negligible amount. This follows from the presence of a finite correlation length in the system that suppresses correlations between distant events exponentially. Hence, the change in the entropy by the deformation should be more or less the same if viewed from the region ABC or BC alone, as A is distant to the deformation. Similarly, the change in the entropy should be the same for the region AC or C alone. This makes $\Delta S_{\text{topo}} = 0$, signifying the independence of topological entropy on the detailed geometric characteristics of the partitioning regions. Of course, this result holds for deformations induced to any single boundary.

We now consider the deformation of the triple point between regions B, C and D, as shown in Figure 9.4(b). To write down the corresponding change in S_{topo} we need the following property. For a pure state the entropy of the reduced density matrix from a bipartition in R and its complement $\bar{R}$ satisfies $S_R = S_{\bar{R}}$. By direct evaluation we have

$$\Delta S_{\text{topo}} = (\Delta S_B - \Delta S_{AB}) + (\Delta S_C - \Delta S_{AC}) + (\Delta S_D - \Delta S_{AD}). \tag{9.13}$$

As the region A is far from the dislocation of the triple point, its appearance in the entropy of the composite regions should not affect the change in the topological entropy. The same reasoning applies to the dislocation of any other triple point of the graph in Figure 9.3. Topological entropy is hence invariant under geometrical distortions of its regions as long as the general topological configuration of these regions remains intact.

9.2.2.2 Universality

Next, we would like to show that topological entropy is not affected when the Hamiltonian of the system is smoothly deformed. This means that S_{topo} can be employed to describe the behaviour of general families of Hamiltonians exhibiting similar topological behaviours. We initially assume that the Hamiltonian is a sum of local terms and that the deformation is restricted to a small region. We require that the deformation not only keeps the system far from any critical point, but also that the correlation length stays small compared to the size of the partition regions given in Figure 9.3 at all times.

The central argument is that any deformation of the Hamiltonian by a localised interaction h,

$$H \to H' = H + h, \tag{9.14}$$

away from the boundaries of the regions, as shown in Figure 9.5, affects the ground state of the system only locally. Hence, it has only a negligible effect on its behaviour near the boundaries and does not affect S_{topo}.

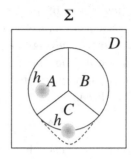

A localised deformation h of the Hamiltonian H is depicted. The deformation can happen either away from the boundaries or on the boundaries. In the latter case we can deform the relevant boundary in order to have h included in a single region.

We now assume that the Hamiltonian H changes in the vicinity of a boundary. Then we can employ the invariance of topological entropy under boundary deformations and move the boundary away from the region where the deformation of the Hamiltonian is to take place. By the previous argument we can smoothly take the system to the new Hamiltonian configuration, H', without changing the value of S_{topo}. As S_{topo} with respect to H' is topologically invariant, we can again move the boundaries without changing the value of the topological entropy. Hence, we can bring the boundary back to its original position. This shows that the topological entropy is a universal quantity with each of its values corresponding to a whole class of Hamiltonians with the same topological order.

9.2.3 Topological entropy and Wilson loops

In this subsection we take a different approach to topological entropy (Levin and Wen, 2006). We show that a change of the entropy S_A from the value $S_A = \alpha|\partial A|$, expected from local correlations alone, to the value $S_A = \alpha|\partial A| - \gamma$ can be attributed to a Wilson loop operator acquiring non-zero expectation value. This loop type of order is revealed by entropic considerations in the following way.

Consider the entanglement entropy of the regions depicted in Figure 9.6. The constant part can be isolated by the following combination of entropies:

$$S_{\text{topo}} = \frac{1}{2}\left[(S_A - S_B) - (S_C - S_D)\right]. \tag{9.15}$$

Substituting (9.7) into (9.15), we find $S_{\text{topo}} = -\gamma$. This relation is exact in the limit where the typical sizes of the involved regions X are all large enough so that $\epsilon(|\partial X|^{-\beta}) \to 0$. Note that A, B, C and D are now different partitions of the same system rather than parts of the same partition as employed in (9.10).

Relation (9.15) gives an intuitive picture for the topological character of γ. As the system is gapped, entanglement entropies acquire, in general, contributions from short-range correlations. The difference $S_A - S_B$ has contributions that come from the upper horizontal part of the region A. Similarly, non-zero contributions to $S_C - S_D$ come from the same

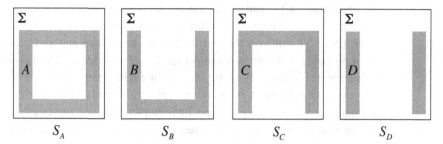

Fig. 9.6   The partitioning of the surface Σ in four different ways, A, B, C and D, with entanglement entropies S_A, S_B, S_C and S_D, respectively. Unlike B, C or D, the region A can support large loop operators that give the non-local γ contribution to the topological entropy, S_{topo}.

upper part. If these entropies only had contributions from local correlations (i.e., if it was $S_X = \alpha|\partial X|$), their difference would go to zero when the respective regions become sufficiently large. The only possible contribution in (9.15) could arise from a non-local operator, like a Wilson loop, that wraps non-trivially around the region A. Such an operator has support on all the regions A, B, C and D of Figure 9.6. Its contribution cannot be present in the entropies of S_B, S_C or S_D, so it cannot be cancelled. A non-zero S_{topo} hence signals a non-zero expectation value of a Wilson loop operator, even if the exact form of this operator is not known.

The above discussion reveals the connection between topological entropy and Wilson loop operators. A non-zero value of S_{topo} means that there should be a certain loop operator with a non-zero expectation value. This remains true for arbitrarily large loop sizes. This type of correlation might seem in stark contrast to the local correlations that one expects to find in a gapped system. The reconciliation comes from the following observation. The exponential decay of correlations of gapped systems, given in (9.2), corresponds to operators $O(\mathbf{r})$ that are local, i.e., defined in a neighbourhood of a given point $\mathbf{r}$. In contrast to this, Wilson loops are non-local operators that have non-trivial action along a whole loop.

A non-zero expectation value of a Wilson loop operator directly implies a non-local structure in the ground state. This is indeed the general approach of the string-net condensation models presented in Levin and Wen (2006). These models consist of a large class of topological models built exactly upon the loop structure of ground states. Nevertheless, the possibility of finding a Wilson loop operator that acquires a non-zero expectation value is a general property of any topological model. Characteristics such as topological degeneracy or anyonic statistics can be expressed in terms of such operators, as we have seen in the previous chapters.

9.3 Example I: Quantum double models

Here we demonstrate that the entanglement entropy of topological models can always be written as

$$S_A = \alpha|\partial A| - \gamma. \tag{9.16}$$

We show this for the particular case of the quantum double models that we encountered in Chapter 5. The approach of Iblisdir *et al.* (2009) generalises the one taken in Hamma *et al.* (2005). It employs the fact that the entanglement entropy measures the entanglement properties between two distinct regions. The Schmidt decomposition of the ground state becomes particularly useful in analysing this bipartition. This decomposition gives a convenient expression of states in terms of their subsystems. So we can straightforwardly evaluate the corresponding entropy. For completeness, we also briefly review the quantum double models with a slightly different approach than the one taken in Chapter 5.

9.3.1 Hamiltonian and its ground state

9.3.1.1 Hamiltonian

Let us consider a square lattice with N links defined on a sphere. The model works in pretty much the same way in an arbitrary planar lattice. We employ a finite group G with $|G|$ elements. The Hilbert space $\mathcal{H} = \{|g\rangle, g \in G\}$ of pairwise orthogonal states $|g\rangle$ is assigned to each link of the lattice. As we have seen in Chapter 5, the lattice Hamiltonian of the quantum double model, $D(G)$, is defined by

$$H = -\sum_v A(v) - \sum_p B(p), \tag{9.17}$$

where the summations run over the vertices v and the plaquettes p of the lattice. The action of the projection operators $A(v)$ and $B(p)$ is given in Figure 9.7. The $A(v)$ operator symmetrises the states neighbouring the v vertex with respect to all group elements of G. The $B(p)$ operator projects the states of the four links of the p plaquette to the state that encodes trivial flux.

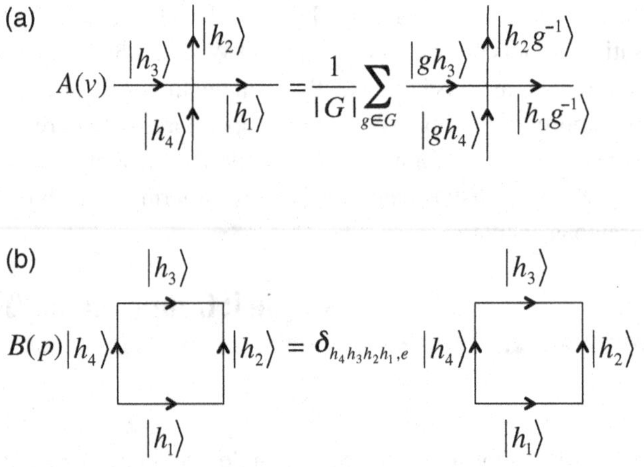

Fig. 9.7 (a) The action of the $A(v)$ operator on the four links neighbouring the vertex v. (b) The action of $B(p)$ on the four links of the plaquette p. The orientation is chosen such that the horizontal links are oriented rightwards and the vertical links are oriented upwards.

9.3.1.2 Ground state

To write down the ground state of the Hamiltonian (9.17) we follow the same approach, which we employed for the toric code in Subsection 5.2.1. We observe that all the $A(v)$'s and $B(p)$'s commute with each other and with themselves for any v and p (see Exercise 5.1). As a result, the ground state can be evaluated as the state that minimises the eigenvalue of each of the $A(v)$ and $B(p)$ terms. For that we first define the reference state $|\,\mathbf{e}\rangle = |\,e...e\rangle$, which assigns the identity group element to every link. We notice that

$$B(p)\,|\,\mathbf{e}\rangle = |\,e\cdots e\rangle = |\,\mathbf{e}\rangle, \tag{9.18}$$

for all p, since $\delta_{h_4h_3h_2h_1,e} = 1$ for $h_1 = h_2 = h_3 = h_4 = e$. Hence, $|\,\mathbf{e}\rangle$ minimises the energy of the $B(p)$ contributions to the Hamiltonian. We then project the $|\,\mathbf{e}\rangle$ state on the eigenstate of the $A(v)$'s with eigenvalue one to minimise the contribution of the $A(v)$ term. The resulting unnormalised ground state is given by

$$|\,\Psi\rangle = \prod_v A(v)\,|\,\mathbf{e}\rangle = \sum_{\mathbf{g}\in\mathbf{G}} R(\mathbf{g})|\mathbf{e}\rangle. \tag{9.19}$$

Here $\mathbf{g} = (g_1,\ldots,g_N)$ belongs in the extended group $\mathbf{G} = G\times\cdots\times G$, which describes any possible rotation of the link states of the lattice, and R is the representation of $\mathbf{G}$ determined by the form of the $A(v)$ operators in Figure 9.7(a). Importantly, $R(\mathbf{g})$ is constructed out of products of $A(v)$ which are elementary loop operators wrapped around vertices.

9.3.2 Topological entropy

Relation (9.19) gives the ground state in terms of equal superpositions of all possible elements $R(\mathbf{g})$ applied to the reference state $|\,\mathbf{e}\rangle$. To understand the properties of $|\,\Psi\rangle$, we first analyse the properties of the group $\mathbf{G}$. This allows us to perform the Schmidt decomposition of the ground state and, thus, to evaluate the entanglement entropy.

9.3.2.1 Schmidt decomposition

Consider a many-particle system split into two subsystems A and B. In the following $\{|\,e_i\rangle, i = 1,\ldots,n\}$ denotes the orthonormal basis of the A part and $\{|f_i\rangle, i = 1,\ldots,m\}$ denotes the orthonormal basis of the B part. A general state of the system can always be written

$$|\,\Psi\rangle = \frac{1}{\sqrt{M}}\sum_{i=1}^{M}\sum_{j=1}^{M} a_{ij}\,|\,e_i\rangle_A \otimes |f_j\rangle_B, \tag{9.20}$$

where $M = \min\{n,m\}$ and where the a_{ij}'s are normalised so that $\sum_{i=1}^{M}\sum_{j=1}^{M}|a_{ij}|^2 = M$. The Schmidt decomposition (Nielsen and Chuang, 2000) asserts that unitaries U and V can be found, for which the state of the system can be written

$$|\Psi\rangle = \frac{1}{\sqrt{M}} \sum_{i=1}^{M} c_i |u_i\rangle_A \otimes |v_i\rangle_B. \tag{9.21}$$

Here $|u_i\rangle = U|e_i\rangle$ and $|v_i\rangle = V|f_i\rangle$ are new orthonormal basis states of A and B, respectively, while the c_i's are real non-negative numbers with $\sum_{i=1}^{M} c_i^2 = M$.

The Schmidt basis in terms of $|u_i\rangle$ and $|v_i\rangle$ makes it easy to calculate reduced density matrices and, from there, the von Neumann entropy of subsystem A. Consider the density matrix of the state $|\Psi\rangle$ given above:

$$\rho = |\Psi\rangle\langle\Psi| = \frac{1}{M} \sum_{i,j=1}^{M} c_i c_j |u_i\rangle_A \langle u_j| \otimes |v_i\rangle_B \langle v_j|. \tag{9.22}$$

The reduced density matrix with respect to region A is then given by

$$\rho_A = \mathrm{tr}_B \rho = \frac{1}{M} \sum_{i,j=1}^{M} c_i c_j |u_i\rangle_A \langle u_j| \otimes \langle v_i|v_i\rangle_B \langle v_j|v_i\rangle = \frac{1}{M} \sum_{i=1}^{M} c_i^2 |u_i\rangle_A \langle u_i|. \tag{9.23}$$

The eigenvalues of the diagonal matrix ρ_A are given by

$$\lambda_i = \frac{c_i^2}{M} \text{ for } i = 1,\dots,M. \tag{9.24}$$

Hence, the von Neumann entropy is given by

$$S(\rho_A) = -\sum_{i=1}^{M} \lambda_i \log \lambda_i = -\frac{1}{M} \sum_{i=1}^{M} c_i^2 \log c_i^2 + \log M. \tag{9.25}$$

In the particular case where $c_i = 1$ for all i, we have $S(\rho_A) = \log M$. This is the case we shall employ in the following.

9.3.2.2 Group decomposition

As the elements of $\mathbf{G}$ can have arbitrary configurations on the two-dimensional lattice, we would like to group them in a simple geometrical way. Consider a region A, defined as a collection of contiguous edges, and its complement B, as shown in Figure 9.8. Ultimately, we want to derive an expression for the entropy of the reduced density matrix ρ_A of the ground state $|\Psi\rangle$ in region A. We can distinguish amongst three types of vertices, those in A, those in B and those touched by edges belonging to A and B. The set of vertices of the latter type is referred to as ∂A, which constitutes the boundary between the two regions. Both regions A and B are taken to be connected and their boundary is taken to always separate links of both regions, i.e., for each vertex $s \in \partial A$, there exist vertices $s' \in A, s'' \in B$ adjacent to s.

As the ground state is stabilised by all the interaction terms of the Hamiltonian (9.17), its properties emerge from the properties of the group $\mathbf{G}$. Thus, we would like to partition the group elements of $\mathbf{G}$ in terms of elements that act only on the links of region A or only on the links of region B. We define two subgroups in $\mathbf{G}$ by $\mathbf{G}_A$ and $\mathbf{G}_B$:

$$\mathbf{G}_A = \{\mathbf{g} \in \mathbf{G}; g_j = e \text{ if } s_j \notin A\}, \quad \mathbf{G}_B = \{\mathbf{g} \in \mathbf{G}; g_j = e \text{ if } s_j \notin B\}. \tag{9.26}$$

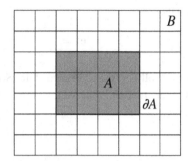

Fig. 9.8   The partitioning of the square lattice into two regions A and B with boundary ∂A. The boundary ∂A is taken along links that lie in-between the links that support the Hilbert spaces of region A and B.

The subgroups $\mathbf{G}_A$ and $\mathbf{G}_B$ act non-trivially only on the links of regions A and B, respectively. In the following we call the elements of $\mathbf{G}_A$ as $\mathbf{g}_A \otimes \mathbb{1}_B$, or $\mathbf{g}_A$ with the identity implied and the elements of $\mathbf{G}_B$ as $\mathbb{1}_A \otimes \mathbf{g}_B$, or just $\mathbf{g}_B$. We now define the quotient group

$$\mathbf{G}_{AB} = \frac{\mathbf{G}}{\mathbf{G}_A \times \mathbf{G}_B}. \tag{9.27}$$

This structure implies that two elements $\mathbf{h}, \mathbf{h}' \in \mathbf{G}_{AB}$ are equal if and only if $h_j = h_j'$ for every vertex $s_j \in \partial A$. In conclusion, a general element $\mathbf{g} \in \mathbf{G}$ can always be written as

$$\mathbf{g} = (\mathbf{g}_A \otimes \mathbf{g}_B)\, \mathbf{h}, \tag{9.28}$$

where $\mathbf{g}_A \in \mathbf{G}_A$, $\mathbf{g}_B \in \mathbf{G}_B$ and $\mathbf{h} \in \mathbf{G}_{AB}$. This means, we achieved a decomposition of $\mathbf{G}$ in terms of the group elements that belong to A, B and their boundary ∂A.

9.3.2.3 Ground state decomposition

The above decomposition of the group according to the bipartition of the system helps us to decompose its ground state. By a straightforward substitution of (9.28) into (9.19), we have

$$|\Psi\rangle = \sum_{\mathbf{h} \in \mathbf{G}_{AB}, \mathbf{g}_X \in \mathbf{G}_X} (\mathbf{g}_A \otimes \mathbf{g}_B)\, \mathbf{h} \,|\, \mathbf{e}\rangle = \mathbf{Q}_A \otimes \mathbf{Q}_B \sum_{\mathbf{h} \in \mathbf{G}_{AB}} \mathbf{h}|\mathbf{e}\rangle, \tag{9.29}$$

where $\mathbf{Q}_X = \sum_{\mathbf{g}_X \in \mathbf{G}_X} \mathbf{g}_X$. The operators $\mathbf{Q}_X$ are local in the X region so they do not change the entanglement properties between regions A and B (Hamma *et al.*, 2005). Similarly to the von Neumann entropy (9.25) of the state (9.21), the only contribution of (9.29) to the entropy comes from the summation part over orthogonal states. There might be inequivalent $\mathbf{h}$ elements in (9.29) that have the same action on the state $|\mathbf{e}\rangle$. In order to find all the independent vectors $\mathbf{h} \,|\, \mathbf{e}\rangle$ with $\mathbf{h} \in \mathbf{G}_{AB}$, we now employ the following analysis. Consider the diagonal subgroup $\mathbf{G}_{\mathrm{d}} \subset \mathbf{G}$ with elements

$$\mathbf{r} = (r, \dots, r), \tag{9.30}$$

where $r \in G$. Then

$$\mathbf{r} \,|\, \mathbf{e}\rangle = \left|\, rer^{-1}...rer^{-1}\right\rangle = |\, \mathbf{e}\rangle\,, \tag{9.31}$$

i.e., the $\mathbf{r}$'s act trivially on the reference state. So, some of the terms in the summation are equal to each other. To avoid that we take the further quotient $\mathbf{G}_{AB}/\mathbf{G}_\mathrm{d}$ and we decompose

$$\mathbf{h} = \mathbf{fr}\,, \tag{9.32}$$

where $\mathbf{f} \in \mathbf{G}_{AB}/\mathbf{G}_\mathrm{d}$ is non-diagonal. As there are in total $|G|$ such $\mathbf{r}$ diagonal elements, we can write

$$|\,\Psi\rangle = \mathbf{Q}_A \otimes \mathbf{Q}_B \sum_{r \in G} \sum_{\mathbf{f} \in \mathbf{G}_{AB}/\mathbf{G}_\mathrm{d}} \mathbf{f}\,\mathbf{r}\,|\,\mathbf{e}\rangle = |G|\mathbf{Q}_A \otimes \mathbf{Q}_B \sum_{\mathbf{f} \in \mathbf{G}_{AB}/\mathbf{G}_\mathrm{d}} \mathbf{f}\,|\,\mathbf{e}\rangle\,, \tag{9.33}$$

where the $\mathbf{f}$ summation runs over the remaining $|\mathbf{G}_{AB}|/|G|$ distinct elements of $\mathbf{G}_{AB}/\mathbf{G}_\mathrm{d}$. All the states $\mathbf{f}\,|\,\mathbf{e}\rangle$ are now independent. So we can apply the Schmidt decomposition (9.21) to the ground state and calculate its entanglement entropy.

9.3.2.4 Entanglement entropy

In order to bring the ground state into the Schmidt decomposition form, and from there calculate $S(\rho_A)$, we need to partition $\mathbf{f}\,|\,\mathbf{e}\rangle$ into two orthogonal basis states. Let $\mathbf{f}$ denote an arbitrary group element of $\mathbf{G}_{AB}/\mathbf{G}_\mathrm{d}$ and let $\mathbf{f} = \bar{\mathbf{f}}_A \otimes \bar{\mathbf{f}}_B$ denote its decomposition into an operator acting on A and an operator acting on B. We then have $\langle \mathbf{e}\,|\,\bar{\mathbf{f}}_X\,|\,\mathbf{e}\rangle = 0$ for both $X = A, B$ as $\mathbf{f} \neq \mathbf{r}$ for all $\mathbf{r} \in \mathbf{G}_\mathrm{d}$ (see Exercise 9.3). Subsequently,

$$|\,\Psi\rangle = |G| \sum_{\mathbf{f} \in \mathbf{G}_{AB}/\mathbf{G}_\mathrm{d}} \left(\mathbf{Q}_A \bar{\mathbf{f}}_A\,|\,\mathbf{e}\rangle_A\right) \otimes \left(\mathbf{Q}_B \bar{\mathbf{f}}_B\,|\,\mathbf{e}\rangle_B\right) \tag{9.34}$$

is actually a Schmidt decomposition of the ground state. This form allows us to appropriately normalise $|\,\Psi\rangle$. Employing (9.25) for $c_i = 1$ for all i, as is the case in (9.34), we can show that the entanglement entropy is of the form

$$S(\rho_A) = \log \frac{|\mathbf{G}_{AB}|}{|G|} = \log |\mathbf{G}_{AB}| - \log |G|, \tag{9.35}$$

where $|\mathbf{G}_{AB}|/|G|$ is the total number of orthogonal states in the Schmidt decomposition (9.34). Here $|G|$ is the order of the group G, i.e., the quantum dimension of the anyonic quantum double model (Brennen *et al.*, 2009). As $|\mathbf{G}_{AB}| = |G|^{|\partial A|}$, where $|\partial A|$ is the number of vertices on the boundary of A, we finally have

$$S(\rho_A) = \log |G|\,(|\partial A| - 1)\,. \tag{9.36}$$

Hence, the topological entropy is $\gamma = \log |G|$. This is a general result that gives the topological entropy for all quantum double models. For the toric code case with quantum dimension $\mathcal{D} = |G| = 2$, we have $\gamma = \log 2$.

Summary

In this chapter we observed that topological behaviour of quantum systems naturally appears in the comprehensive description of entropy. The most general form of the entanglement entropy of a gapped two-dimensional system prepared in a pure state, with respect to a bipartition A and B, is necessarily given by

$$S(\rho_A) = \alpha |\partial A| - \gamma. \tag{9.37}$$

In addition to the area law, $\alpha |\partial A|$, dictated from short-range correlations we need to include a constant term γ. A non-zero value of this term reveals non-local correlations present in the topological model. In particular, we have

$$\gamma = \log \mathcal{D}, \tag{9.38}$$

where $\mathcal{D}$ is the total quantum dimension of the topological model. As is common to science, this discovery came about by considering the entropy of a specific example, the toric code (Hamma *et al.*, 2005), where the peculiar presence of a constant term was observed. The actual definition of topological entanglement entropy $S_{\text{topo}} = -\gamma$ in terms of physical observables and the study of its properties was presented later by Kitaev and Preskill (2006) and by Levin and Wen (2006). Measuring topological order in an experiment is expected to be a difficult task, which reflects the high complexity involved in detecting non-local order.

The topological entropy is a diagnostic tool that identifies if a system is topologically ordered or not. If topological order is identified, then we know that the system can support anyons. We can then start envisioning schemes for performing topological quantum computation. As γ is a function of the total quantum dimension, $\mathcal{D}$, of a system, it cannot distinguish successfully between different anyonic models that have the same $\mathcal{D}$. Direct study of anyonic properties could eventually provide this information.

Exercises

9.1 Demonstrate relations (9.11), (9.12), (9.13) and (9.15).

9.2 Find the most general form of S_{topo} defined with the partitions given in Figure 9.3 that is invariant under arbitrary deformations of the boundaries. [*Hint*: To solve that take an arbitrary linear combination of the entropies and determine the coefficients by demanding invariance of S_{topo} with respect to boundary and triple point deformations (Iblisdir *et al.*, 2009).]

9.3   Consider an element $\mathbf{f}$ that belongs in $\mathbf{G}_{AB}/\mathbf{G}_{\text{d}}$ defined in (9.27) and (9.30). This element acts non-trivially on both A and B. Suppose $\mathbf{f} = \bar{\mathbf{f}}_A \otimes \bar{\mathbf{f}}_B$. Show that $\langle \mathbf{e} | \bar{\mathbf{f}}_X | \mathbf{e} \rangle = 0$ for $X = A$ or B, where $| \mathbf{e} \rangle$ is defined in (9.18). [*Hint*: Suppose it is false. See also Iblisdir *et al.* (2009).]

10

In the previous chapters we introduced anyons and their properties, we presented how to perform topological quantum computation and studied several examples of topological models. There is a wide variety of research topics concerned with topological quantum computation. Among the many open questions, two have a singular importance. The first natural question is: which physical systems can support non-Abelian anyons? Realising non-Abelian anyons in the laboratory is of fundamental and practical interest. Such exotic statistical behaviour has not yet been encountered in nature. The physical realisation of non-Abelian anyons would be the first step towards the identification of a technological platform for the realisation of topological quantum computation. The second question concerns the efficiency of topological systems in combating errors. It has been proven that the effect of coherent environmental errors in the form of local Hamiltonian perturbations can be suppressed efficiently without degrading the topologically encoded information (Bravyi *et al.*, 2010). Nevertheless, there is no mechanism that can protect topological order from incoherent probabilistic errors. Topological systems nevertheless constitute a rich and versatile medium that allows imaginative proposals to be developed (Chesi *et al.*, 2010; Hamma *et al.*, 2009).

Regarding the first question, we can identify two main categories of physical proposals for the realisation of two-dimensional topological systems: systems that are defined on the continuum and discrete systems defined on a lattice. It is natural to ask, which are the most promising architectures to realise in the laboratory? Undoubtedly, fractional quantum Hall liquids are so far the most studied topological systems. They comprise a two-dimensional cloud of electrons in the presence of a strong perpendicular magnetic field. There is a big variety of topological phases that arise as a function of the magnetic field and of the density of electrons (Duan and Guo-jun, 1992; MacDonald, 1990; Nayak *et al.*, 2008; Prange and Girvin, 1990). The most striking characteristic is that in some of these phases the particles appear to have fractionalised charge in units of the electron charge. This was first verified experimentally by Tsui *et al.*, (1982). Subsequently, the charge fractionalisation was explained theoretically by Laughlin (1983) in terms of Abelian anyons which emerge as quasiparticles in this highly correlated system.

The fractional quantum Hall effect is currently the only experimental setting we have that promises the realisation of non-Abelian anyons. The theoretical treatment of this system is limited due to the interactions between the electrons that make the problem too hard to solve analytically or numerically. So experiments are crucial to determine if a given fractional quantum Hall setting can support non-Abelian anyons. The current focus of research is on the filling fraction $\nu = 5/2$. This filling factor is expected to support the non-Abelian anyons which correspond to the SU(2) level-2 Chern–Simons theory. These anyons are similar to the Ising ones, so they are not universal for quantum computation. Nevertheless, inspiring proposals exist to rectify this (Bravyi, 2006; Bravyi and Kitaev, 2005; Das Sarma *et al.*, 2005). Filling fraction $\nu = 12/5$ is expected to correspond to the SU(2) level-3 non-Abelian anyons. As these are equivalent to the Fibonacci anyons they can support universal quantum computation.

An important diagnostic tool of fractional quantum Hall liquids is the implementation of interference experiments. These can be used to detect the anyonic character of quasiparticles. By particular configurations of the liquid sample we could isolate and detect Abelian (Chamon *et al.*, 1997) and non-Abelian (Bonderson *et al.*, 2006, 2007; Chung and Stone, 2006; Stern and Halperin, 2006) statistical properties. Together with the braiding operations, such measurements can be employed to perform topological quantum computation (Das Sarma *et al.*, 2005). Presently, it is nevertheless not known how to realise the transport of quasiparticles in fractional quantum Hall liquids. Hence, we are not yet able to directly implement the braiding of anyons. The measurement of anyons can be employed instead in order to realise a version of one-way quantum computation with static anyons (Bonderson *et al.*, 2008, 2009). Using this scheme, measurements of the topological charge can be used to generate the braiding transformations used in topological quantum computation without the need to physically transport anyons. Being able to bypass the anyonic transport overcomes a crucial obstacle in the implementation of topological quantum computation with fractional quantum Hall liquids.

Other electron systems are the *p*-wave superconductors that can support fractionally charged vortices with anyonic statistics (Ivanov, 2001; Read and Green, 2000). They have been shown to be equivalent to Kitaev's honeycomb lattice model that we studied in Chapter 6 (Chen and Nussinov, 2008; Yu and Wang, 2008). Recently, topological insulators were discovered, which have similar properties as the integer and the fractional quantum Hall effect. Interestingly, they do not require the presence of an external magnetic field to acquire topological properties (Kane and Mele, 2006). These materials are the focus of a rapidly developing field, both theoretically and experimentally (Hasan and Kane, 2010).

As an alternative to the continuous models, one can consider two-dimensional lattice models. There, one aims to engineer detailed systems with a known anyonic content instead of searching for them in nature, as we do with the fractional quantum Hall effect. For example, the quantum double models in Chapter 5 have been proposed to be realised with Josephson junctions (Doucot *et al.*, 2004). First experiments already realised the required four-spin interactions of the toric code (Gladchenko *et al.*, 2009). Moreover, the quantum simulation of Abelian anyonic statistics has been performed by encoding the toric code states in the polarisation states of four (Pachos *et al.*, 2009) or six photons (Lu *et al.*, 2009).

In addition, Levin and Wen proposed a family of topological models called the string-net models that include the quantum doubles as special cases (Levin and Wen, 2005). This generalised class provides a versatile laboratory to theoretically probe anyons. A general drawback of the string-net or quantum double models is that they need interactions between more than two particles that are hard to engineer in the laboratory. In contrast to this, Kitaev's honeycomb lattice model supports non-Abelian anyons and requires only two-spin interactions (Kitaev, 2006). A proposal for its realisation with cold atoms was given by

Duan *et al.* (2003). Moreover, it has been proposed by Micheli *et al.* (2006) to implement the honeycomb lattice model with polar molecules (Brennen *et al.*, 2007; Büchler *et al.*, 2007).

The second question we posed concerns the stability of topological order under environmental perturbations. Topological order is stable at zero temperature, when the system is subject to local, weak and time-independent perturbations (Bravyi *et al.*, 2010). Such perturbations can be of the form of erroneous interaction terms added to the Hamiltonian of the system (Pastawski *et al.*, 2009). The stability proof is based on generic properties of topological systems, such as the non-local character of their quantum correlations. This resilience of topological models was initially conjectured by Kitaev (2003). It actually motivated him to employ anyons for quantum computation. The stability of topological order can be translated directly to resilience of topological quantum computation. Indeed, not only the ground state is topologically protected, but also the anyonic excitations as well as their statistical behaviour. The proven stability translates to the resilience of topological quantum computation against zero-temperature perturbations. Moreover, it has been shown that the toric code is stable against detectable probabilistic loss of edges, with a tolerated loss rate as high as 50% (Stace *et al.*, 2009). A stability study of the toric code under a quantum quench is given in Tsomokos *et al.* (2009).

But can topological order survive at non-zero temperatures? The effect of finite temperatures on topological systems such as the quantum doubles has been studied with the help of topological entanglement entropy. It was shown that for fixed temperature the topological entropy vanishes as the size of the system is increased (Castelnovo and Chamon, 2007; Iblisdir *et al.*, 2009). Indeed, a finite temperature induces errors to the system with a finite probability, in the form of unwanted excitations. In other words, topological systems do not have an intrinsic mechanism which protects them against probabilistic errors. From the explicit form of the thermal states given in (3.9), we can deduce that increasing the energy gap of the system decreases the probability of such errors occurring, but the energy gap cannot eliminate them completely.

From the above discussion it is clear that topological systems by themselves are not able to protect topologically encoded information from temperature errors. Are there any modifications we can perform in order to achieve this goal? The aim is to safely store quantum information in a system, which is subject to finite temperature, for arbitrarily long times and without performing continuous quantum error correction. Recently, two inspiring schemes appeared that address this problem within the context of topological systems. Hamma *et al.* (2009) decorated the toric code with a scalar field that couples to anyons, thus creating long-range attractive interactions between them. When temperature errors in the form of anyonic excitations occur, then the attractive interaction causes them to annihilate. Still, topological order remains intact as the ground state does not sense the presence of the scalar field. If such a system could be designed, it would be characterised by a finite critical temperature below which information could be reliably stored. An alternative scheme was presented by Chesi *et al.* (2010). It employs repulsive long-range interactions between the anyons of the toric code. They showed that such a quantum memory is protected against temperature fluctuations as it energetically penalises the generation of anyonic errors in the system. These are just two examples that suggest how to combat finite temperature errors.

It is an exciting possibility to imagine what types of physical effects one could employ in topological systems in order to improve the performance of quantum memories.

Beyond these two questions presented above there is a variety of problems that need to be addressed. The better we understand the forms quantum matter can take, the better we can encode and manipulate quantum information for technological applications, such as performing error-free quantum computation. The theoretical and experimental investigation of topological systems moreover provides a platform to study the physics of anyons in their own right. We can probe and manipulate anyonic quasiparticles and study their physics, such as transport phenomena or critical behaviour, without the need to resort to the microscopic properties of the underlying topological system. This opens a wealth of possibilities, where anyons can be the ingredients for fundamental research and for technological applications.

References

Aguado, M., Brennen, G. K., Verstraete, F. and Cirac, J. I. 2008. *Phys. Rev. Lett.* **101**, 260501.

Aharonov, D. 2007. *SIAM J. Comput.* **37**, 166.

Aharonov, D., Jones, V. and Landau, Z. 2009. *Algorithmica* **55**, 395.

Aharonov, Y. and Bohm, D. 1959. *Phys. Rev.* **115**, 485.

Alexander, J. W. 1923. *Proc. Natl. Acad. Sci. USA* **9**, 93.

Alexander, J. W. and Briggs, G. B. 1926. *Ann. Math.* (2) **28**, 562.

Anderson, P. W. 1958. *Phys. Rev.* **109**, 1492–505.

Arovas, D., Schrieffer, J. R. and Wilczek, F. 1984. *Phys. Rev. Lett.* **53**, 722.

Avron, J. E., Osadchy, D. and Seiler, R. 2003. *Physics Today* **56**, 38.

Avron, J. E., Seiler, R. and Simon, B. 1983. *Phys. Rev. Lett.* **51**, 51.

Bais, F. A., van Driel, P. and de Wild Propitius, M. 1992. *Phys. Lett. B* **280**, 63.

Baraban, M., Zikos, G., Bonesteel, N. and Simon, S. H. 2009. *Phys. Rev. Lett.* **103**, 076801.

Bell, J. S. 1966. *Rev. Mod. Phys.* **38**, 447.

Bennett, C. H. 1982. *Int. J. Theor. Phys.* **21**, 905.

Berry, M. V. 1984. *Proc. Roy. Soc., Ser. A* **392**, 45.

Bohm, A., Mostafazadeh, A., Koizumi, H., Niu, Q. and Zwanziger, J. 2003. *The Geometric Phase in Quantum Systems*. New York: Springer-Verlag.

Bolukbasi, A. T. and Vala, J. 2011. *Non-Abelian Berry Phase Calculations in the Kitaev Honeycomb Model*, arXiv:1103.3061.

Bonderson, P., Kitaev, A. and Shtengel, K. 2006. *Phys. Rev. Lett.* **96**, 016803.

Bonderson, P. 2007. *Non-Abelian Anyons and Interferometry*, PhD thesis.

Bonderson, P., Shtengel, K. and Slingerland, J. K. 2007. *Phys. Rev. Lett.* **98**, 070401.

Bonderson, P., Freedman, M. and Nayak, C. 2008. *Phys. Rev. Lett.* **101**, 010501.

Bonderson, P., Freedman, M. and Nayak, C. 2009. *Ann. Phys.* **324**, 787.

Born, M. and Fock V. A. 1928. *Zeitschrift für Physik* **51**, 165–80.

Bose, S. N. 1924. *Zeitschrift für Physik* **26**, 178–81.

Bravyi, S. and Kitaev, A. 2005. *Phys. Rev. A* **71**, 022316.

Bravyi, S. 2006. *Phys. Rev. A* **73**, 042313.

Bravyi, S., Poulin, D. and Terhal, B. 2009. *Phys. Rev. Lett.* **104**, 050503.

Bravyi, S., Hastings, M. and Michalakis, S. 2010. *J. Math. Phys.* **51**, 093512.

Brell, C. G., Flammia, S. T., Bartlett, S. D. and Doherty, A. C. 2011. *New J. Phys.* **13**, 053039.

Brennen, G. K., Micheli, A. and Zoller, P. 2007. *New J. Phys.* **9**, 138.

Brennen, G. K., Aguado, M. and Cirac, J. I. 2009. *New J. Phys.* **11**, 053009.

Brennen, G. K., Iblisdir, S., Pachos, J. K. and Slingerland, J. K. 2009. *New J. Phys.* **11**, 103023.

Broda, B. 1990. *Mod. Phys. Lett. A* **5**, 2747.

Büchler, H. P., Demler, E., Lukin, M., Micheli, A., Prokof'ev, N., Pupillo, G. and Zoller, P. 2007. *Phys. Rev. Lett.* **98**, 060404.

Burrello, M., Xu, H., Mussardo, G. and Wan, X. 2010. *Phys. Rev. Lett.* **104**, 160502.

Camino, F. E., Zhou, W. and Goldman, V. J. 2007. *Phys. Rev. Lett.* **98**, 076805; 2005. *Phys. Rev. B* **72**, 075342.

Castagnoli, G. and Rasetti, M. 1993. *Int. J. Mod. Phys.* **32**, 2335.

Castelnovo, C. and Chamon, C. 2007. *Phys. Rev. B* **76**, 184442.

Castelnovo, C. and Chamon, C. 2008. *Phys. Rev. B* **78**, 155120.

Chamon, C., Freed, D. E., Kivelson, S. A., Sondhi, S. L. and Wen, X. G. 1997. *Phys. Rev. B* **55**, 2331.

Chamon, C., Jackiw, R., Nishida, Y., Pi, S.-Y. and Santos, L. 2010. *Phys. Rev. B* **81**, 224515.

Chen, H.-D. and Nussinov, Z. 2008. *J. Phys. A -Math. Theor.* **41**, 7.

Cheng, M., Lutchyn, R. M., Galitski, V. and Das Sarma, S. 2009. *Phys. Rev. Lett.* **103**, 107001.

Chern, S. S. 1944. *Ann. Math.* **45**, 747.

Chesi, S. Roethlisberger, B. and Loss, D. 2010. *Phys. Rev. A* **82**, 022305.

Chung, S.-B. and Stone, M. 2006. *Phys. Rev. B* **73**, 245311.

Chung, S.-B. and Stone, M. 2007. *J. Phys. A* **40**, 4923.

Chung, S.-B., Yao, H., Hughes, T. L. and Kim, E.-A. 2010. *Phys. Rev. B* **81**, 060403(R).

Das Sarma, S., Freedman, M. and Nayak, C. 2005. *Phys. Rev. Lett.* **94**, 166802.

Dennis, E., Kitaev, A., Landahl, A. and Preskill, J. 2002. *J. Math. Phys.* **43**, 4452.

Deser, S., Jackiw, R. and Templeton, S. 1982. *Phys. Rev. Lett.* **48**, 975.

Deutsch, D. 1985. *Proc. Roy. Soc. London, Ser. A* **400**, 97.

Dirac, P. A. M. 1926. *Proc. Roy. Soc., Ser. A* **112**, 661–77.

Doucot, B., Ioffe, L. B. and Vidal, J. 2004. *Phys. Rev. B* **69**, 214501.

Duan, F. and Guo-jun, J. 1992. *New Perspective on Condensed Matter Physics*. Shanghai: Shanghai Scientific & Technical Publishers.

Duan, L.-M., Cirac, J. I. and Zoller, P. 2001. *Science* **292**, 1695.

Duan, L.-M., Demler, E. and Lukin, M. D. 2003. *Phys. Rev. Lett.* **91**, 090402.

Dunne, G. V. 1998. *Aspects of Chern-Simons Theory*, Les Houches Lectures.

Einstein, A. 1924. *Sitzungsberichte der Preussischen Akademie der Wissenschaften, Physik-Mathematik*, 261–7.

Einstein, A., Podolsky, B. and Rosen, N. 1935. *Phys. Rev.* **47**, 777–80.

Elitzur, S., Moore, G., Schwimmer, A. and Seiberg, D. 1989. *Nucl. Phys. B* **326**, 108.

Farhi, E., Goldstone, J., Gutmann, S., Lapan, J., Lundgren, A. and Preda, D. 2001. *Science* **292**, 472.

Fermi, E. 1926. *Rend. Lincei* **3**, 145–9.

Feynman, R. P., Leighton, R. B. and Sands, M. 1963. *The Feynman Lectures on Physics* Vol III. New York: Addison-Wesley, Chapter 1.

Feynman, R. P. 1965. *The Feynman Lectures on Physics*, Vol. 3. New York: Addison-Wesley, pp. 1–8.

Feynman, R. 1982. *Int. J. Theor. Phys.* **21**, 467–88.

Finkelstein, D. and Rubinstein, J. 1968. *J. Math. Phys.* **9**, 1762.

Freedman, M., Larsen, M. and Wang, Z. 2002a. *Commun. Math. Phys.* **228**, 177–99.

Freedman, M., Larsen, M. and Wang, Z. 2002b. *Commun. Math. Phys.* **227**, 605–22.

Freedman, M., Kitaev, A. and Wang, Z. 2002c. *Commun. Math. Phys.* **227**, 587–603.

Freedman, M., Kitaev, A., Larsen, M. J. and Wang, Z. 2003. *Bull. Amer. Math. Soc.* **40**, 31.

Froehlich, J., Studer, U. M. and Thiran, E. J. 1997. *Statist. Phys.* **86**, 821.

Fu, L., Kane, C. L. and Mele, E. J., 2007. *Phys. Rev. Lett.* **98**, 106803.

Fulton, W. and Harris, J. 1991. *Representation Theory: A First Course*. Berlin: Springer-Verlag.

Furukawa, S. and Misguich, G. 2007. *Phys. Rev. B* **75**, 214407.

Gasiorowicz, S. 1996. *Quantum Physics*. New York: John Wiley and Sons.

Georgiev, L. S. 2006. *Phys. Rev. B* **74**, 235112.

Gladchenko, S., Olaya, D., Dupont-Ferrier, E., Doucot, B., Ioffe, L. B. and Gershenson, M. E. 2009. *Nature Phys.* **5**, 48–53.

Gottesman, D. 1997. *Stabilizer Codes and Quantum Error Correction*, Caltech PhD thesis, quant-ph/9705052.

Gottesman, D. 1998. *Phys. Rev. A* **57**, 127.

Grover, L. K. 1996. *28th Annual ACM Symposium on the Theory of Computing*, p. 212.

Guadagnini, E., Martinellini, M. and Mintchev, M. 1990. *Nucl. Phys. B* **336**, 581; 1990. *Nucl. Phys. B (Proc. Suppl.)* **18B**, 121.

Guadagnini, E. 1993. *The Link Invariants of the Chern–Simons Field Theory*. Berlin: Walter de Gruyter & Co.

Gurarie, V. and Radzihovsky, L. 2007. *Phys. Rev. B* **75**, 212509.

Hamma, A., Zanardi, P. and Wen, X. -G. 2005. *Phys. Rev. B* **72**, 035307.

Hamma, A., Ionicioiu, R. and Zanardi, P. 2005. *Phys. Lett. A* **337**, 22.

Hamma, A., Zhang, W., Haas, S. and Lidar, D. A. 2008. *Phys. Rev. B* **77**, 155111.

Hamma, A., Castelnovo, C. and Chamon, C. 2009. *Phys. Rev. B* **79**, 245122.

Haque, M., Zozulya, O. and Schoutens, K. 2007. *Phys. Rev. Lett.* **98**, 060401.

Hardy, L. 2001. *Quantum Theory From Five Reasonable Axioms*, quant-ph/0101012.

Hasan, M. Z. and Kane, C. L. 2010. *Rev. Mod. Phys.* **82**, 3045.

Hastings, M. B. and Michalakis, S. 2009. *Quantization of Hall Conductance for Interacting Electrons Without Averaging Assumptions*, arXiv:0911.4706.

Hatfield, B. 1992. *Quantum Field Theory of Point Particles and Strings*. New York: Perseus Books Group.

Iblisdir, S., Pèrez-Garcìa, D., Aguado, M. and Pachos, J. K. 2009. *Phys. Rev. B* **79**, 134303; 2010. *Nucl. Phys. B* **829**, 401.

Ivanov, D. A. 2001. *Phys. Rev. Lett.* **86**, 268.

Jackiw, R. and Rossi, P. 1981. *Nucl. Phys. B* **190**, 681.

Jackson, J. D. 1975. *Classical Electrodynamics*, 2nd edn. Singapore: Wiley Eastern.

Jaeger, F., Vertigan, D. L. and Welsh, D. J. 1990. *A. Math. Proc. Cambridge Phil. Soc.* **108**, 35.

Jaeger, F., Vertigan, D. L., and Welsh, D. J. 1990. *A. Math. Proc. Cambridge Philos. Soc.* **108**, 35–53.

Jiang, H. -C., Gu, Z. -C., Qi, X. -L. and Trebst, S. 2011. *Phys. Rev. B* **83**, 245104.

Jones, V. F. R. 1985. *Bull. Amer. Math. Soc.* **12**, 103.

Jones, V. 2005. notes at http://www.math.berkeley.edu/~vfr/jones.pdf.

Kane, C. L. and Mele, E. J., 2006. *Science* **314**, 1692.

Kargarian, M. and Fiete, G. A. 2010. *Phys. Rev. B* **82**, 085106.

Kauffman, L. H. 1991. *Knots and Physics*. Singapore: World Scientific.

Kauffman, L. H. and Lomanaco, S. 2006. *q-deformed spin networks, knot polynomials and anyonic topological quantum computation*, quant-ph/0606114.

Kells, G., Bolukbasi, A. T., Lahtinen, V., Slingerland, J. K., Pachos, J. K. and Vala, J. 2008. *Phys. Rev. Lett.* **101**, 24.

Kells, G., Slingerland, J. K. and Vala, J. 2009. *Phys. Rev. B* **80**, 125415.

Kells, G., Mehta, D., Slingerland, J. K. and Vala, J. 2010. *Phys. Rev. B* **81**, 104429.

Kitaev, A. 1997. *Russ. Math. Surv.* **52**, 1191.

Kitaev, A. 2000. *Unpaired Majorana fermions in quantum wires*, cond-mat/0010440.

Kitaev, A. 2003. *Ann. Phys.* **303**, 2.

Kitaev, A. 2006. *Ann. Phys.* **321**, 2.

Kitaev, A. and Preskill, J. 2006. *Phys. Rev. Lett.* **96**, 110404.

Landau, L. D. and Lifshitz, E. M. 1977. *Quantum Mechanics: Nonrelativistic Theory*. Oxford: Pergamon Press.

Laughlin, R. 1981. *Phys. Rev. B* **23**, 5632.

Laughlin, R. 1983. *Phys. Rev. Lett.* **50**, 1395.

Lahtinen, V., Kells, G., Stitt, T., Vala, J. and Pachos, J. K. 2008. *Ann. Phys.* **323**, 9.

Lahtinen, V. and Pachos, J. K. 2009. *New J. Phys.* **11**, 093027.

Lahtinen, V. 2011. *New J. Phys.* **13**, 075009.

Lee, D. -H., Zhang, G. -M. and Xiang, T. 2007. *Phys. Rev. Lett.* **99**, 196805.

Leinaas, J. M. and Myrheim, J. 1977. *Nuovo Cimento B* **37**, 1.

Levin, M. A. and Wen, X. -G. 2005. *Phys. Rev. B* **71**, 045110.

Levin, M. A. and Wen, X. -G. 2006. *Phys. Rev. Lett.* **96**, 110405.

Lieb, E. H. 1994. *Phys. Rev. Lett.* **73**, 2158.

Lu, C. -Y., Gao, W. -B., Gühne, O., Zhou, X. -Q., Chen, Z. -B. and Pan, J. -W. 2009. *Phys. Rev. Lett.* **102**, 030502.

MacDonald, A. H. (ed.), 1990. *Quantum Hall Effect: A Perspective*. Dordrecht: Kluwer Academic Publishers.

MacLane, S. 1998. *Categories for the Working Mathematician*, 2nd edn. Graduate Texts in Mathematics. New York: Springer-Verlag.

Majorana, E. 1937. *Nuovo Cimento* **5**, 171–84.

Messiah, A. 1962. *Quantum Mechanics*. Amsterdam: North Holland.

Micheli, A., Brennen, G. K. and Zoller, P. 2006. *Nature Phys.* **2**, 341.

Miller, J. B., Radu, I. P., Zumbÿhl, D. M., Levenson-Falk, E. M., Kastner, M. A., Marcus, C. M., Pfeiffer, L. N. and West, K. W. 2007. *Nature Phys.* **3**, 561.

Mochon, C. 2004. *Phys. Rev. A* **69**, 032306.

Moore, G. and Read, N. 1991. *Nucl. Phys. B* **360**, 362.

Nakahara, M. 2003. *Geometry, Topology, and Physics,* 2nd edn. Oxford: Taylor and Francis.

Nayak, C., Simon, S. H., Stern, A., Freedman, M. and Das Sarma, S. 2008. *Rev. Mod. Phys.* **80**, 1083.

Nechaev, S. 1996. *Statistics of Knots and Entangled Random Walks.* Singapore: World Scientific.

Nielsen, M. A. and Chuang, I. L. 2000. *Quantum Computation and Quantum Information.* Cambridge: Cambridge University Press.

NIST. 2006. *Fine Structure Constant,* The NIST Reference on Constants, Units, and Uncertainty.

Niu, Q. and Thouless, D. J. 1987. *Phys. Rev. B* **35**, 2188.

Nussinov, Z. and Ortiz, G. 2008. *Phys. Rev. B* **77**, 064302.

Nussinov, Z. and Ortiz, G. 2009. *Proc. Natl. Acad. Sci. USA* **106**, 16944.

Pachos, J. K. and Zanardi, P. 2001. *Int. J. Mod. Phys. B* **15**, 1257.

Pachos, J. K. 2002. *Contemp. Math.* **305**, 245.

Pachos, J. K. 2007. *Ann. Phys.* **322**, 1254.

Pachos, J. K., Wieczorek, W., Schmid, C., Kiesel, N., Pohlner, R. and Weinfurter, H. 2009. *New J. Phys.* **11**, 083010.

Papanikolaou, S., Raman, K. S. and Fradkin, E. 2007. *Phys. Rev. B* **76**, 224421.

Pastawski, F., Kay, A., Schuch, N. and Cirac, I. 2010. *Quant. Inform. Comput.* **10**, 580.

Pauli, W. 1940. *Phys. Rev.* **58**, 716.

Peshkin, M. and Tonomura, A. 1989. *The Aharonov–Bohm Effect,* Lecture Notes in Physics, 340. Berlin: Springer-Verlag.

Polyakov, A. M. 1988. *Mod. Phys. Lett. A* **3**, 325.

Prange, R. and Girvin, S. M. (eds). 1990. *The Quantum Hall Effect.* New York: Springer-Verlag.

Preskill, J. 2004. Lecture Notes for Physics 219: Quantum Computation. http://www.theory.caltech.edu/~preskill/.

Rauch, H., Zeilinger, A., Badurek, G., Wilfing, A., Bauspiess, W. and Bonse, U. 1975. *Phys. Lett. A* **54**, 425.

Raussendorf, R. and Briegel, H. -J. 2001. *Phys. Rev. Lett.* **86**, 5188.

Raussendorf, R., Harrington, J. and Goyal, K. 2007. *New J. Phys.* **9**, 199.

Read, N. and Green, D. 2000. *Phys. Rev. B* **61**, 10267.

Read, N. 2009. *Phys. Rev. B* **79**, 045308.

Reidemeister, K. 1926. *Abh. Math. Sem. Univ. Hamburg* **5**, 24.

Roland, J. and Cerf, N. J. 2002. *Phys. Rev. A* **65**, 042308.

Rowell, E., Stong, R. and Wang, Z. 2009. *Commun. Math. Phys.* **292**, 343.

Schmidt, K. P., Dusuel, S. and Vidal, J. 2008. *Phys. Rev. Lett.* **100**, 057208.

Shor, P. 1995. *Phys. Rev. A* **52**, 2493.

Shor, P. 1997. *SIAM J. Sci. Statist. Comput.* **26**, 1484.

Simon, S. H., Bonesteel, N. E., Freedman, M. H., Petrovic, N. and Hormozi, L. 2006. *Phys. Rev. Lett.* **96**, 070503.

Slater, J. C. 1929. *Phys. Rev.* **34**, 10.

Stace, T. M., Barrett, S. D. and Doherty, A. C. 2009. *Phys. Rev. Lett.* **102**, 200501.

Steane, A. M. 1996. *Phys. Rev. Lett.* **77**, 793.

Stern, A., von Oppen, F. and Mariani, E. 2004. *Phys. Rev B* **70**, 205338.

Stern, A. and Halperin, B. I. 2006. *Phys. Rev. Lett.* **96**, 016802.

Stern, A. 2008. *Ann. Phys.* **323**, 204–49.

Stone, M. and Chung, S.-B. 2003. *Phys. Rev. B* **73**, 014505.

Sundermeyer, K. 1982. *Constrained Dynamics*. Berlin: Springer-Verlag.

Symanzik, K. 1983. *Nucl. Phys. B* **190**, 1.

Thouless, D. J., Kohmoto, M., Nightingale, M. P. and den Nijs, M. 1982. *Phys. Rev. Lett.* **49**, 405.

Tserkovnyak, Y. and Simon, S. H. 2003. *Phys. Rev. Lett.* **90**, 016802.

Tsomokos, D. I., Hamma, A., Zhang, W., Haas, S. and Fazio, R. 2009. *Phys. Rev. A* **80**, 060302(R).

Tsui, D. C., Störmer, H. L. and Gossard, A. C. 1982. *Phys. Rev. B* **25**, 1405–7.

Tsui, D. C., Störmer, H. L. and Gossard, A. C. 1982. *Phys. Rev. Lett.* **48**, 1559.

Turaev, V. G. 1994. *Quantum invariants of knots and 3 manifolds*. de Gruyter Studies in Mathematics, Vol. 18, Berlin: Walter de Gruyter & Co.

Turing, A. M. 1937. *Proc. London Math. Soc.* **42**, 230–65.

Verlinde, E. 1988. *Nucl. Phys. B* **300**, 360.

Vidal, J., Schmidt, K. P. and Dusuel, S. 2008. *Phys. Rev. B* **78**, 245121.

Volovik, G. E. 2003. *The Universe in a Helium Droplet*. Oxford: Oxford University Press.

von Klitzing, K., Dorda, G. and Pepper, M. 1980. *Phys. Rev. Lett.* **45**, 494.

Wallace, P. R. 1947. *Phys. Rev.* **71**, 622.

Wang, Z. 2010. *Topological Quantum Computation*, CBMS Regional Conference Series in Mathematics.

Weimer, H., Müller, M., Lesanovsky, I., Zoller, P. and Büchler, H. P. 2010. *Nature Phys.* **6**, 382.

Weinberg, S. 1995. *The Quantum Theory of Fields*. Cambridge: Cambridge University Press.

Wen, X. G. 1995. *Adv. Phys.* **44**, 405.

Wilczek, F. 1982. *Phys. Rev. Lett.* **49**, 957.

Wilczek, F. and Zee, A. 1984. *Phys. Rev. Lett.* **52**, 2111.

Wilczek, F. 2009. *Nature Phys.*, **5**, 614.

Willett, R. L., Pfeiffer, L. N. and West, K. W. 2009. *Proc. Natl. Acad. Sci. USA* **106**, 8853.

Witten, E. 1989. *Commun. Math. Phys.* **121**, 351.

Wootters, W. K. and Zurek, W. H. 1982. *Nature* **299**, 802–3.

Wootton, J. R., Lahtinen, V. and Pachos, J. K. 2009. *LNCS* **5906**, 56.

Wootton, J. R. 2011. *Increasing the stability of the topological entanglement entropy* arXiv:1103.2878.

Yang, C. N. 1970. *Phys. Rev. D* **1**, 2360.

Yao, H. and Kivelson, S. A. 2007. *Phys. Rev. Lett.* **99**, 247203.

Yao, H., Zhang, S. -C. and Kivelson, S. A. 2009. *Phys. Rev Lett.* **102**, 217202.

Yao, H. and Qi, X. -L. 2010. *Phys. Rev. Lett.* **105**, 080501.

Yu, Y. and Wang, Z. 2008. *Europhys. Lett.* **84**, 57002.

Zanardi, P. and Rasetti, M. 1999. *Phys. Lett. A* **264**, 94.

Index